Signals and Communication Technology

For further volumes:
http://www.springer.com/series/4748

Parth H. Pathak · Rudra Dutta

Designing for Network
and Service Continuity
in Wireless Mesh
Networks

 Springer

Parth H. Pathak
Department of Computer Science
North Carolina State University
Raleigh
NC, USA

Rudra Dutta
Department of Computer Science
North Carolina State University
Raleigh
NC, USA

ISSN 1860-4862
ISBN 978-1-4614-4626-2   ISBN 978-1-4614-4627-9   (eBook)
DOI 10.1007/978-1-4614-4627-9
Springer New York Heidelberg Dordrecht London

Library of Congress Control Number: 2012949378

Printed on acid-free paper

Springer is part of Springer Science+Business Media (www.springer.com)

To my late grandparents
Mr. and Mrs. Brahmbhatt, and
To my beloved mother and father
—who each unknowingly started shaping
this years ago

P.H.P.

To my father
Sambhu Nath Dutta, and
To my father-in-law,
Subas Chandra Bose,
—who each exemplified in their lives the
best of professional engineering, and inspired
others

R.D.

Preface

The idea for this book, and the book itself, grew out of our research on mesh networks at North Carolina State University, over the period 2010–2012, during which the first author was doing doctoral research under the guidance of the second. Our research addressed various design topics in wireless multihop networking, especially the mesh paradigm. Much of the research focused on the impact of power control on mesh design, and using power control as a design tool. While we were focusing on performance issues, in keeping with contemporary research, the question slowly formed in our minds as to whether these issues were indeed the most pressing ones.

Conversations with colleagues in both academia and industry led us to believe that understanding and improving the issues of predictability of performance, and tolerance of (and robustness under) wireless disruptions and other faults, are appropriate issues for mesh design research to address at this time. Indeed, they are perhaps among the most critical, in that they point the path to providing continuity characteristics of services delivered using mesh networks—of practical importance in affecting the real-world deployment and adoption of such networks. The latter part of our research collaboration focused on these issues, some of which is represented in this book in summary.

However, it also became clear that this research area is not as well explored as it could be, and as we expect it to be in the near future. We had achieved a certain understanding of the background and issues involved in such research; while we plan to continue research in this area, it seemed appropriate to contribute our understanding to the community, in the shape of this book. We hope it will provide input to some researchers working in this area, and perhaps help, in a small way, inform their research. Although we believe we have done a reasonably comprehensive job, we consider this book to be far from the last word; we are grateful if it makes no more than an effective beginning for some researchers, in pursuing this newly emerging area.

Despite our best efforts, some factual errors may have escaped us, or we may have inadvertently misinterpreted or misrepresented some literature; we sincerely apologize for any such, and would not only welcome but value corrections from definitive sources. Obviously, such errors are our own, and not those of the sources we cite.

We are grateful to some colleagues at NC State, particularly Dr. Mihail L. Sichitiu, for illuminating discussions. We also acknowledge the US Army Research Office, which, under grants W911NF-08-1-0105 and W911NF-09-1-0341, supported us over most of this period. While these grants did not fund this effort directly, they enabled us to build the CentMesh outdoor wireless mesh testbed at NCSU, which provided us with invaluable practical insight and experience in mesh design issues, without which this book would have been far less informed.

Raleigh, NC, USA, May 2012   Parth H. Pathak
                                                                  Rudra Dutta

Contents

Abbreviations

3GPP	The 3rd Generation Partnership Project
CDMA	Code Division Multiple Access
FCC	Federal Communications Commission
IEEE	Institution of Electrical and Electronics Engineers
ILP	Integer-Linear Programming
IMT-Advanced	International Mobile Telecommunications—Advanced
IP	Internet Protocol
ISM	Industrial, Scientific, and Medical radio bands
LP	Linear Programming
LTE	Long Term Evolution
MAC	Medium Access Control
PHY	Physical layer of the Open System Interconnect model
SINR	Signal to Interference and Noise Ratio
SNR	Signal to Noise Ratio
TCP	Transmission Control Protocol
U-NII	Unlicensed National Information Infrastructure
UHF	Ultra High Frequency (0.3–3 GHz)
VHF	Very High Frequency (30–300 MHz)
WLAN	Wireless Local Area Networks
WiFi	A common name for the IEEE 802.11 standards
WiMAX	A common name for the IEEE 802.16 standards

Chapter 1
Introduction

1.1 Perspective

The paradigm of Wireless Mesh Network (WMN) was one of several which arose from the earlier concept of ad-hoc wireless networks, which was also the precursor of other specific domains of multi-hop wireless networking such as sensor networks or vehicular networks. The mesh paradigm was seen as one characterized by static routing nodes, with available electrical power, thus enabling high transmission power levels and increased geographic span; possibly integrating other, mobile or low-power, nodes, and possibly serving client nodes which do not participate in routing or other network functions. The paradigm soon emerged as a potentially suitable solution for metropolitan area networks, providing *last few miles last few hops* connectivity. There are various attractive qualities of this paradigm, which include low-cost deployment, robustness and applicability to brownfield scenarios where installation of new ground fiber is infeasible, but Internet access retrofit is required.

The mesh paradigm inherits useful characteristics from both the ad-hoc networking paradigm and the traditional wired infrastructure paradigm. In a way, the mesh paradigm might be seen to provide a middle ground between extreme regimes of wireless use. On the one hand, we can conceive of networks in which wireless links provide only last-hop links, connecting to a wired backbone—both WiFi and 2G/3G networks may be seen as such cases, though the difference in technology makes for a very different span in the two cases. At the other extreme, an approach of exclusively wireless links to form a network with an arbitrary topology out of a significant number of nodes and an extended span has become mainstream for multi-hop wireless paradigms such as sensor, mobile ad-hoc, or tactical networks—such networks may not even connect to the wired backbone, or connect through a gateway as an isolated and separately managed network. The mesh network paradigm is more general in that it does not decree a specific pattern or number of wired network connections, type of clients, or even mobility. The path from a particular mesh node (or client node served by a mesh node) may have many wireless links to its nearest wired

P. H. Pathak and R. Dutta, *Designing for Network and Service Continuity in Wireless Mesh Networks*, Signals and Communication Technology, DOI: 10.1007/978-1-4614-4627-9_1, © Springer Science+Business Media New York 2013

gateway, or one, or none. This makes the mesh network suitable for gaining general understanding of various networks scenarios (though more detailed understanding still invites a consideration of specific scenarios), since the other paradigms could be viewed as special cases of the mesh. In this way, the mesh paradigm provides a conceptual convergence point. The trend towards more flexibility in more rigidly structured paradigms, such as the introduction of femtocell or mircocell approaches in cellular telephony, or the use of WiMax (2007) as backhaul, show that the diverse scenarios posed in mesh research, allowed by the broadness of the mesh paradigm, are likely to be useful in reflecting emerging real-world problems.

This breadth also results in two fundamental benefits of the generality of the WMN paradigm; the potential for easy deployment and affordable cost. This also explains why wireless multihop testbeds are generally considered "mesh testbeds". Such testbeds, and many real deployments, often use WiFi technology for individual links, though the IEEE 802.11 standard was hardly designed with such use in mind. This by no means restricts the WMNs' applicability to other standards, but cheap availability of 802.11 hardware has naturally motivated this focus. Because the 802.11 software stack was originally designed forinfrastructure WLANs, various modifications are necessary when using it in WMNs. Researchers are actively investigating these modifications, and the majority of efforts are directed towards design of better link layer and channel access protocols. Meanwhile, other standards like WiMax (2007) and 3G/4G continue to emerge and mature, and the knowledge gained by research and development of WMNs over 802.11 is likely to be very useful in the future in these diverse contexts.

After its original inception, the concept of mesh networking soon attained a comparatively stable form, commonly understood and agreed upon by the community. This paradigm has been competently described, and research literature on the topic surveyed, by various previous work, such as Akyildiz et al. (2005). Research efforts focused on many topics, both analysis and design oriented. Much of the early research focused on traditionally important issues of performance, such as throughput, delay, guaranteed performance and QoS, etc. In the last five to ten years, there has been a tremendous quickening of research interest in this area, with increased understanding of the design and deployment of such networks. One of the things that has become clear, through experimental academic testbeds and real-life deployments, is that the design problems that have been studied in isolation, such as routing, channel assignment, power control, topology control, etc., are so closely linked through the reality of wireless interference, that joint approaches to design are likely to provide much better results in practice. From the point of view of the practitioner, this is unfortunate; joint design methods are notoriously complicated, and difficult to translate into practice and maintain. In addition, different joint design studies typically make their own assumptions about the integrated framework in which design may be carried out, and there is no commonly accepted converged framework.

Both results of theoretical studies and simulations, and testbed experience, have shown that, despite the comparatively static nature and high transmission power usually assumed for WMNs for Internet retrofit purposes, the ever-present uncertainty characteristics of the wireless medium result in a network that has significantly less predictability of service characteristics than has typically come to be expected of carrier-grade access networks. In practical terms, some observers have suggested this as being the primary cause slowing down practical adoption and deployment of mesh-based solutions (Doyle 2010; Lawson 2008). Thus research in the design of WMNs has increasingly focused on issues such as availability, reliability, and service continuity, and attempts to quantify them, or characterize a network in these terms. The joint design approaches mentioned earlier have proved most valuable in such pursuits. These can now be considered the most current and pressing issues in WMN research.

In the rest of this chapter, we introduce a commonly accepted view of WMNs, and also include some discussion on testbeds, as preliminaries. As indicated above, a mesh testbed requires careful design and meticulous consideration of various hardware/software aspects (Robinson et al. 2005), without which performance evaluation done using the testbed can be misleading or even erroneous. Accordingly, as the deployment of testbeds proceeded both to verify research and for commercial ventures, the need for research that considered design in realistic (i.e., joint) terms became more sharply felt; in turn, mesh testbeds became further necessary to verify the results of such research. Experience with such testbeds also provided a clear view of the need of research in mesh survivability. We see this interaction as the main driver of research in joint design, and later survivability design, in mesh networks.

1.2 WMN Architecture, Characteristics and Benefits

A wireless mesh network consists of wireless mesh routers and wired/wireless clients (See Fig. 1.1). Wireless mesh routers communicate in multi-hop fashion forming a relatively stable network. Clients connect to these routers using a wireless or a wired link. In the most common form of WMNs, every router performs relaying of data for other mesh routers (a typical ad-hoc networking paradigm), and certain mesh routers also have the additional capability of being Internet gateways. Such gateway routers often have a wired link which carries the traffic between the mesh routers and the Internet. This general form of WMNs can be visualized as an integration of two planes where the access plane provides connectivity to the clients while the forwarding plane relays traffic between the mesh routers. This design has become more and more popular due to the increasing usage of multiple radios in mesh routers and virtual wireless interfacing techniques.

Though WMNs inherit almost all characteristics of the more general ad-hoc network paradigm, such as decentralized design, distributed communications etc., there are a few differences. Unlike energy-constrained ad-hoc networks, mesh routers have no limitations regarding energy consumption. Also, the pattern of traffic between these routers is assumed to be fairly stable over time, more akin to typical access or campus networks, unlike sensor or tactical wireless networks. For this reason, WMN nodes can also have stable forwarding and routing roles, like more traditional

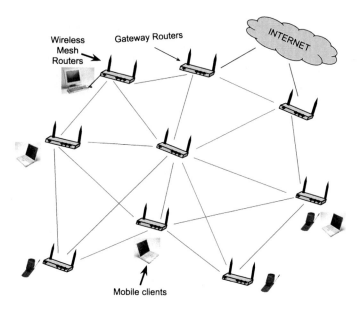

Fig. 1.1 Wireless mesh network architecture—mesh routers, mesh clients and gateway nodes

infrastructure networks. In contrast, when WMNs are deployed for the purpose of short-term mission specific communication, they often act more as a tradition Mobile Ad-hoc Network (MANET). Here, the majority of the traffic flows between mesh routers (not always to the gateways as in previous case) and even clients may communicate with each other directly. This kind of architecture is referred to as a hybrid mesh (Akyildiz et al. 2005) and is one of the promising and emerging vision for the future of WMNs.

There can be pre-planned (usually centrally controlled) as well as comparatively unstructured and incremental deployment of nodes in WMNs. In the recent past, there have been many attempts to design community wireless networks using unstructured deployment of WMNs. In such Wireless Community Networks (WCNs) (Efstathiou et al. 2006), users own the mesh routers and participate in the network to facilitate access to other users for mutual benefit. In developed areas, the fundamental objective of such an unplanned deployment/expansion is to develop an Internet connectivity blanket for anywhere, anytime connectivity (Antoniadis et al. 2008). Also, WMNs deployment has been proposed as reliable and affordable access networks in underdeveloped regions. Here, the aim is to design a network as a low-cost access initiative (often by Internet Service Providers) to aid the development of communities. WMNs benefit from incremental expansion because their robustness and coverage increases as more and more mesh routers are added. Finally, we have already referred to the generality of the mesh paradigm, which allows them to be instantiated quickly with relatively cheap equipment, while studying issues which can be of benefit in designing mesh networks with costlier equipment and larger scope.

1.3 Experimental Testbeds, Real-World Deployments

The attractiveness of the mesh paradigm that we have referred to above have motivated researchers to actually realize such instantiations in studying these networks, and also motivated entrepreneurs to attempt trials deployments.

Simulation based studies of wireless ad-hoc networks have been long conducted and it is known that there is a significant gap between the actual measured performance and simulation results. In the last few years, increasingly cheaper and more accessible technology has allowed researchers to undertake actual testbed based evaluation of protocols. This has lead to research and development of a plethora of mesh testbeds. However, the development of such testbeds also made clear for the first time the critical importance of jointly considering traditionally isolated design problems, because the testbed designer has to make some decisions, if only by default, about the issues that are not of central interest to the research problem at hand. In simulation, it might be feasible to study the relative performance of two particular routing algorithms without making any reference to the medium access approach underneath, but an actual testbed has to use some actual MAC. Moreover, the answer to the comparative performance question may well change depending on what MAC is used—or even details in its configuration, such as the carrier sense threshold of 802.11. Such testbeds thus spurred the quickened interest and explosive growth of the joint design research area that this survey is focused on, and in turn provide the proving ground for such research. The study of joint design in WMNs is thus also, in part, a study of research issues in WMN testbeds. Below we provide only a very brief overview to motivate our discussion on joint design; a full survey of mesh testbeds is outside the scope of this paper and merits a separate discussion.

Examples of such testbeds include MIT Roofnet, CUWIN (2007) at Urbana-Champaign, MeshNet (2007) at UC Santa Barbara, WiseNet (2007) and CentMesh (2007) at North Carolina State University CentMesh (2007), Purdue-Mesh (2007), Broadband Wireless Networking BWN (2007) lab at Georgia Tech, SMesh (Amir et al. 2006) etc. Some testbeds like ORBIT (2007) at Rutgers, and Emulab (2007) at Utah provide flexible platform to other researchers who can test their methodology or protocols on them. Such efforts have given rise to many open source implementations of protocols, device drivers and network applications. Several research efforts are directed towards making community based mesh networks more and more self-organizing and cooperative (MCL 2007) where every participant contributes to the network resources.

Mesh testbeds nodes are typically small single board embedded computers like Soekris (2007) boards, medium capacity machines like VIA (2007) EPIA mini-ITX motherboards or high capacity desktops. When using off-the-shelf hardware, wireless cards using Atheros 802.11 chipsets are often used due to their open source driver support like Madwifi (2007)and recently Ath5k (2007) and Ath9k (2007). Though testbed experimentations result in precise evaluation, they are often time-consuming, costly and inflexible. To overcome such issues, scaled-down, smaller transmission range versions of actual testbeds such as ScaleMesh (ElRakabawy et al. 2008) and

Fig. 1.2 Community wireless mesh network for Internet access

IvyNet (Su and Gross 2008) can also be used. Sometimes a combination of simulation, emulation and real-world testbed experiments are used (Nordström et al. 2007) or testbeds are deployed with advanced operating system virtualization techniques (Zimmermann et al. 2006, 2007) to improve the testbed control and management.

There is a diverse range of application scenarios for wireless mesh network deployment; this is another issue which significantly affects the perceived performance of various isolated design approaches. The fundamental objective of mesh deployment has been low-cost Internet access. Mesh networks deployed in communities spanning small or medium sized areas can be a very good business model for ISPs to provide Internet access (See Fig. 1.2). TFA Rice mesh (Camp et al. 2006), Heraklion Mesh (Delakis et al. 2008), Google-Meraki mesh are a few of the examples of such deployments. With recent awareness about using alternate sources of energy, many of the wireless mesh routers are also designed to run with solar energy and rechargeable batteries (SolarMESH 2007). This will certainly give rise to mesh deployments in near future where mesh routers running on solar energy can be fixed on apartment roofs or light poles, forming a mesh in neighborhood areas. Mesh networks can also serve the purpose of temporary infrastructure in disaster and emergency situations. Various control systems such as public area surveillance can also be operated using WMNs. Other applications considered for WMNs include remote medical care (Takahashi et al. 2007), traffic control system (Lan et al. 2007), public services

(Bernardi et al. 2008), integration with sensor monitoring systems (Wang et al. 2008; Wu et al. 2006). Considering this plethora of applications, many vendors have started providing mesh based network solution for broadband Internet access. Strix (2007) systems, Cisco (2007) systems, Firetide (2007), Meraki (2007a,b) MeshDynamics (2007), BelAir (2007), Tropos (2007) and PacketHop (2007) are some examples of commercial WMN vendors.

1.4 Overview

In this book, we have attempted to present an integrated view into the topic of designing WMNs with a view to survivability issues, and the current state of the art. Since the issue of joint design is central to such design, we have included some material on this in Chap. 5 [we refer the reader interested in more detail to our previous more thorough survey on this topic (Pathak and Dutta 2010)] before discussing the survivabilty related research in Chap. 6. In order to ensure self-sufficiency, and provide a pathway to these design discussions for those readers previously unfamiliar with WMN design, we have briefly discussed enabling technologies in Chap. 2, and basic design issues in Chaps. 3 and 4. Obviously, we needed to assume a baseline familiarity with wireless networking topics; a full treatment of each topic from first principles would destroy the coherence and focus of this work, as well as growing intractably large. As an example, we discuss some details of the IEEE standard 802.11n, which is both comparatively new and also increasingly relevant for WMNs; but we assume familiarity with the 802.11 a/b/g protocols, which have been widely used for many years. In a similar fashion, we have discussed some research areas and literature in more depth than others, trying to keep in mind both the relative relevance of these areas to the focus of our work, and the relative familiarity the reader is likely to have for the various areas.

References

Akyildiz IF, Wang X, Wang W (2005) Wireless mesh networks: a survey. Comput Netw ISDN Syst 47(4):445–487. doi:10.1016/j.comnet.2004.12.001

Amir Y, Danilov C, Hilsdale M, Musăloiu-Elefteri R, Rivera N (2006) Fast handoff for seamless wireless mesh networks. In: Proceedings of the 4th international conference on mobile systems, applications and services (MobiSys '06). ACM, New York, NY, USA, pp 83–95. doi:10.1145/1134680.1134690

Antoniadis P, Le Grand B, Satsiou A, Tassiulas L, Aguiar R, Barraca J, Sargento S (2008) Community building over neighborhood wireless mesh networks. IEEE Technol Soc Mag 27(1):48–56. doi:10.1109/MTS.2008.918950

Ath5k (2007) Ath5k. http://madwifi-project.org/wiki/About/ath5k

Ath9k (2007) Ath9k. http://linuxwireless.org/en/users/Drivers/Atheros

BelAir (2007) BelAir networks. http://www.belairnetworks.com

Bernardi G, Buneman P, Marina MK (2008) Tegola tiered mesh network testbed in rural scotland. In: Proceedings of the 2008 ACM workshop on wireless networks and systems for developing regions (WiNS-DR '08). ACM, New York, NY, USA, pp 9–16: doi:10.1145/1410064.1410067

BWN (2007) BWN. http://www.ece.gatech.edu/research/labs/bwn/mesh/index.html

Camp J, Robinson J, Steger C, Knightly E (2006) Measurement driven deployment of a two-tier urban mesh access network. In: Proceedings of the 4th international conference on Mobile systems, applications and services (MobiSys '06). ACM, New York, NY, USA, pp 96–109. doi:10.1145/1134680.1134691

CentMesh (2007) The centennial wireless mesh project. http://centmesh.csc.ncsu.edu/

Cisco (2007) Cisco systems. www.cisco.com/web/go/outdoorwirelessnetworks

CUWIN (2007) CUWIN Community wireless networks. http://cuwireless.net

Delakis M, Mathioudakis K, Petroulakis N, Siris VA (2008) Experiences and investigations with Heraklion mesh: an experimental metropolitan multi-radio mesh network. In: Proceedings of the 4th international conference on testbeds and research infrastructures for the development of networks and communities (TridentCom '08), pp 1–6.

Doyle M (2010) Wimax woes in high-rise chicago. Windy Citizen Tech

Efstathiou EC, Frangoudis PA, Polyzos GC (2006) Stimulating participation in wireless community networks. In: Proceedings of 25th IEEE international conference on computer communications (INFOCOM 2006), pp 1–13. doi:10.1109/INFOCOM.2006.320

ElRakabawy S, Frohn S, Lindemann C (2008) Scalemesh: a scalable dual-radio wireless mesh testbed. In: Proceedings of 5th IEEE annual communications society conference on sensor, mesh and Ad Hoc communications and networks workshops (SECON Workshops '08), pp 1–6. doi:10.1109/SAHCNW.2008.21

Emulab (2007) Emulab.http://www.emulab.net/

Firetide (2007) Firetide Inc. http://www.firetide.com

Lan Kc, Wang Z, Hassan M, Moors T, Berriman R, Libman L, Ott M, Landfeldt B, Zaidi Z (2007) Experiences in deploying a wireless mesh network testbed for traffic control. SIGCOMM Comput Commun Rev 37(5):17–28. doi:10.1145/1290168.1290171

Lawson S (2008) Backhaul woes slow sprint's wimax rollout. PCWorld Magazine

Madwifi (2007) MadWifi. http://madwifi-project.org/

MCL (2007) Microsoft connectivity layer. http://research.microsoft.com/en-us/projects/mesh/

Meraki (2007a) Meraki Inc. http://meraki.com

Meraki (2007b) Meraki Mesh. http://sf.meraki.com/map

MeshDynamics (2007) MeshDynamics. http://www.meshdynamics.com

MeshNet (2007) MeshNet. http://moment.cs.ucsb.edu/meshnet/

MIT-Roofnet (2007) MIT Roofnet. http://pdos.csail.mit.edu/roofnet/doku.php

Nordström E, Gunningberg P, Rohner C, Wibling O (2007) Evaluating wireless multi-hop networks using a combination of simulation, emulation, and real world experiments. In: Proceedings of the 1st international workshop on system evaluation for mobile platforms (MobiEval '07). ACM, New York, NY, USA, pp 29–34. doi:10.1145/1247721.1247728

ORBIT (2007) ORBIT Lab.http://www.orbit-lab.org/

PacketHop (2007) PacketHop Inc.http://www.packethop.com

Pathak P, Dutta R (2010) A survey of network design problems and joint design approaches in wireless mesh networks. IEEE Commun Surv Tutor 13(99):1–33. doi:10.1109/SURV.2011.060710.00062

Purdue-Mesh (2007) Purdue Mesh. http://engineering.purdue.edu/MESH

Robinson J, Papagiannaki K, Diot C, Guo X, Krishnamurthy L (2005) Experimenting with a multi-radio mesh networking testbed. In: 1st workshop on wireless network measurements (Winmee). Riva del Garda, Italy

Soekris (2007) Soekris engineering. http://www.soekris.com

SolarMESH (2007) SolarMesh.http://owl.mcmaster.ca/~todd/SolarMESH

Strix (2007) Strix systems.http://www.strixsystems.com

Su Y, Gross T (2008) Validation of a miniaturized wireless network testbed. In: Proceedings of the third ACM international workshop on wireless network testbeds, experimental evaluation and characterization (WiNTECH '08). ACM, New York, NY, USA, pp 25–32. doi:10.1145/1410077. 1410084

Takahashi Y, Owada Y, Okada H, Mase K (2007) A wireless mesh network testbed in rural mountain areas. In: Proceedings of the second ACM international workshop on wireless network testbeds, experimental evaluation and characterization (WinTECH '07). ACM, New York, NY, USA, pp 91–92. doi:10.1145/1287767.1287785

Tropos (2007) Tropos networks.http://www.tropos.com

VIA (2007) VIA technologies. http://www.via.com.tw

Wang KC, Venkatesh G, Pradhananga S, Lokala S, Carter S, Isenhower J, Vaughn J (2008) Building wireless mesh networks in forests: antenna direction, transmit power, and vegetation effects on network performance. In: Proceedings of the third ACM international workshop on Wireless network testbeds, experimental evaluation and characterization (WiNTECH '08). ACM, New York, NY, USA, pp 97–98. doi:10.1145/1410077.1410097

WiMax (2007) WiMax. http://www.wimaxforum.org/

WiseNet (2007) WiseNet. http://netsrv.csc.ncsu.edu/twiki/bin/view/Main/Wisenet

Wu D, Gupta D, Liese S, Mohapatra P (2006) Qurinet: quail ridge natural reserve wireless mesh network. In: Proceedings of the 1st international workshop on wireless network testbeds, experimental evaluation and characterization (WiNTECH '06). ACM, New York, NY, USA, pp 109–110. doi:10.1145/1160987.1161015

Zimmermann A, Günes M, Wenig M, Ritzerfeld J, Meis U (2006) Architecture of the hybrid mcg-mesh testbed. In: Proceedings of the 1st international workshop on wireless network testbeds, experimental evaluation and characterization (WiNTECH '06). ACM, New York, NY, USA, pp 88–89. doi:10.1145/1160987.1161004

Zimmermann A, Günes M, Wenig M, Meis U, Ritzerfeld J (2007) How to study wireless mesh networks: a hybrid testbed approach. In: Proceedings of the 21st international conference on advanced networking and applications, IEEE computer society (AINA '07). Washington, DC, USA, pp 853–860. doi:10.1109/AINA.2007.77

Chapter 2
Mesh Enabling Technology

2.1 IEEE 802.11 and Its Amendments

Ease of deployment and affordable cost are two main reasons behind the increasing popularity of wireless mesh networks. Compared to other alternatives of wireless access networks such as cellular networks, wireless mesh networks can potentially provide carrier-grade Internet services at a lower capital expenditure (CAPEX) and operational expenditure (OPEX). IEEE 802.11 technology has been the key in enabling low-cost wireless multi-hoping due to its support of ad-hoc networking. Because of this reason, many current wireless mesh network deployments are based on IEEE 802.11 standards. This by no means restricts the applicability of WMNs to other standards; but cheaper cost, flexibility and higher availability of 802.11 hardware and software are the factors that have most motivated the growth.

IEEE 802.11 a/b/g are most commonly used wireless technology standards for mesh networking. Since 802.11 a and g standards can provide higher data rates (upto 54 Mbps), they have become more popular in recent WMN deployments. A typical two-tier mesh network consists of an *access* tier and a *backhaul* tier. The access tier provides connectivity between mesh routing nodes and their clients, while the backhaul tier consists of interconnections among the mesh routers. In order to avoid interference between the two tiers, the access tier typically operates in 802.11 b/g mode while the backhaul tier operates in 802.11 a mode. This mitigates the inter-tier interference because 802.11 a uses the 5 GHz ISM band and 802.11 b/g use 2.4 GHz.

Even though most of the WMN deployments use IEEE 802.11 a/b/g standards, an additional amendment is proposed in the form of IEEE 802.11s standard. The motivation behind the design and development of 802.11s is that the a/b/g standards were not designed for multi-hop communications. Although the 802.11 a/b/g have been reasonably well leveraged for mesh, they were originally designed to operate in infrastructure WLANs. In order to address the issues of coordinated medium access, 802.11s proposes *Mesh Deterministic Access* (MDA), built on the idea that contention for access to the medium should be separated as much as possible from the actual medium utilization. We will discuss IEEE 802.11s MAC in more detail in a later chapter. The major difference between 802.11s and the other 802.11 standards is

P. H. Pathak and R. Dutta, *Designing for Network and Service Continuity in Wireless Mesh Networks*, Signals and Communication Technology, DOI: 10.1007/978-1-4614-4627-9_2, © Springer Science+Business Media New York 2013

how mesh nodes access the medium. Functionality of other layers in 802.11s remain more or less similar; e.g. 802.11s uses similar PHY layer as a/g for carrying the traffic.

2.2 Multiple Input Multiple Output (MIMO) Based IEEE 802.11

The performance gains of utilizing multiple antennas in wireless networks have been long explained by seminal works of Foschini and Gans (1998), Telatar (1999). The systems in which multiple antennas are used at the wireless receiver and transmitter are referred to as Multiple-Input Multiple-Output (MIMO) (Fig. 2.1), as opposed to systems where receiver and transmitter each have a single antenna—Single-Input Single-Output (SISO). MIMO technology has been employed in many current mobile standards such as LTE, WiMAX and 802.11n.

MIMO technology can increase the throughput of a wireless channel, but (more importantly for our present context), it can improve the consistency and predictability of the channel, as we describe below. For this reason, it is an important technology for service continuity issues. Here, we restrict out attention to the IEEE 802.11n standard due to its applicability in wireless mesh networks. We also discuss multi-hop WiMAX networks in the Sect. 2.3.

Foschini and Gans (1998) and Telatar (1999) showed for the first time that capacity increases linearly when an additional pair of antennas are added at link end-points. This is an especially important result since the capacity gain is achieved even when

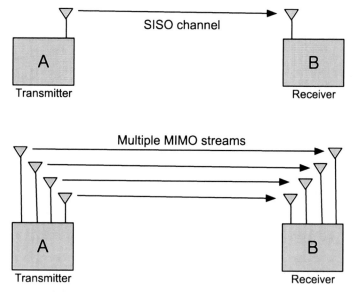

Fig. 2.1 SISO and MIMO: a transmitter and receiver can communicate with a single antenna pair, or with multiple antenna pairs constituting a MIMO link

both receiver and transmitter are tuned on the same channel; thus it represents a more effective utilization on the same spectrum, not the use of additional antennas to access additional spectrum. The gain is attributed to the creation of independent spatial paths between pairs of antennas, which allows significantly more information to be exchanged at the same time. Previously 802.11a/g have also employed multiple antennas for capacity gain. This is different from latter MIMO systems such as 802.11n; in the former, the best signal out of multiple antennas is chosen, while the latter allows parallel processing of data from all antennas. The theoretical achievable throughput of 802.11n is 600 Mbps as opposed to 54 Mbps attainable in 802.11a/g. As shown by Halperin et al. (2010), the increase is due to multiple antennas, increase in channel width, and link layer frame aggregation. The increase of data rate using multiple antennas can be leveraged by the backhaul links of wireless mesh networks, which typically experience stable high traffic demand.

Similar to 802.11a/g, 802.11n uses Orthogonal Frequency Division Multiplexing (OFDM). When operating in non-HT (High Throughput) mode, a A 20 MHz channel is divided into 56 subcarriers (out of which 52 subcarriers are usable) when operating in High Throughput (HT) mode, or 52 subcarriers (48 usable) when operating in non-HT mode.[1] Similarly, a 40 MHz channel is divided into 114 subcarriers where 108 carriers can be used for transmission. This is shown in Fig. 2.2. The spectral mask of a 40 MHz channel is shown in Fig. 2.3. Using this spectral mask, the 5.4 GHz U-NII band can be divided into 5 orthogonal channels (Fig. 2.4). Due to the larger spread, multiple orthogonal channels of 40 MHz can not be obtained in 2.4 GHz spectrum where 802.11g devices are largely located (Fig. 2.5). We will see later how their coexistence can create prohibitive throughput decrements in 802.11n links. Due to this reason, the 2.4 GHz spectrum is largely unsuitable for 802.11n operations.

There are multiple reasons behind performance increase of 802.11n as compared to 802.11a/g. First, 40 MHz channels provide higher link throughput. 802.11n is also effective in combating multi-path fading (Judd et al. 2008), because such fading effects are largely frequency specific, and when sufficient redundancy is added in subcarrier information, it is possible to decode the information even if multiple consecutive subcarriers are affected due to multi-path fading. Since such techniques are already employed in 802.11a/g, the added advantage of 802.11n comes due to multiple antennas that can allow spatial diversity. Details of gains due to spatial diversity and frame aggregation are listed in the next subsection.

Table 2.1 lists the data rates achievable by the 802.11n standard. Modulation and Coding Scheme (MCS) is a number derived based on combinations of modulation, coding rate, guard period size, channel width and number of spatial streams. The guard period is the time between two consecutive transmissions of symbols, necessary in order to adjust for delayed receptions due to multi-path effects. Finally, a number of spatial streams is established in each case, between that number of antenna pairs in parallel.

---

[1] Some subcarriers are used as pilots for dynamic calibration, and are not usable for data transmission.

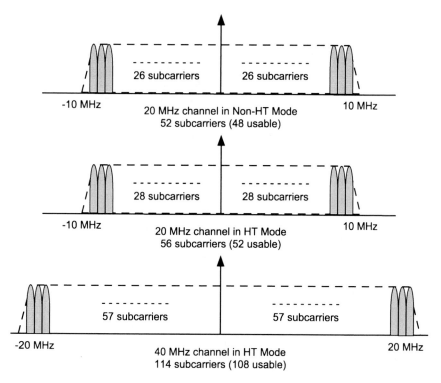

Fig. 2.2 Increased number of subcarriers allows larger and more reliable information exchange in 802.11n

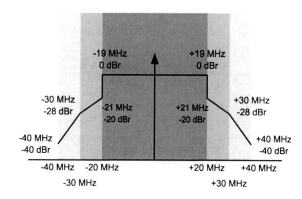

Fig. 2.3 Spectral mask of 40 MHz channel allowed in 802.11n

2.2.1 Spatial Diversity

A set of techniques are applied to the receiver and transmitter in order to leverage the multiple signals received by multiple antennas.

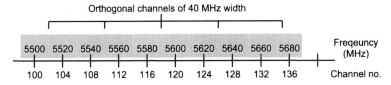

Fig. 2.4   U-NII Spectrum band (5.450–5.725 GHz) divided into 5 channels of 40 MHz

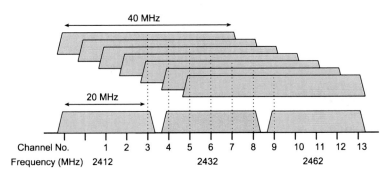

Fig. 2.5   40 MHz channels are not suitable for 2.4 GHz band

2.2.1.1   Receiver Diversity

In order to understand receiver diversity, let us first consider a sample two-node network shown in Fig. 2.6. In the network, both nodes are equipped with 3 antennas. Node A (transmitter) only uses one antenna to transmit the signal. Node B (receiver) uses all three antennas to receive the transmitted signal. Receiver diversity techniques are used to combine the received signals of each antenna in order to constructively determine the transmitted information. Following Halperin et al. (2010), we discuss two methods of receiver diversity.

1. Strongest-signal-only (SSO): The antenna that receives the strongest signal will be considered for frame reception. The method is simple and is in fact helpful in reliability since it provides a choice of potentially better signal. On the other hand, the received signals at the other antennas are simply wasted.
2. Maximal Ratio Combining (MRC): Signals are superimposed with each other such that they are in the same phase. This allows constructive addition of the signals, which is likely to be better than the SSO signal. Further, before addition of the signals, they can be weighted using their SNR values to avoid the impact of noise from weaker signals on MRC. Most of the current 802.11n implementations use MRC for receiver diversity.

Halperin et al. (2010) present results regarding the performance of different receiver diversity methods when a 1×3 topology similar to Fig. 2.6 is implemented using commodity hardware in indoor environment. As can be expected, their results verify that the signals received by individual antennas suffer multipath fading in certain subcarriers. On the other hand, when MRC is used, the resultant signal strength

Table 2.1 Achievable 802.11n data rates using various modulations, coding rates, number of spatial streams and guard intervals

MCS index	Type	Coding rate	Spatial streams	Data rate (Mbps) with 20 MHz CH		Data rate (Mbps) with 40 MHz CH	
				800 ns	400 ns (SGI)	800 ns	400 ns (SGI)
0	BPSK	1/2	1	6.50	7.20	13.50	15.00
1	QPSK	1/2	1	13.00	14.40	27.00	30.00
2	QPSK	3/4	1	19.50	21.70	40.50	45.00
3	16-QAM	1/2	1	26.00	28.90	54.00	60.00
4	16-QAM	3/4	1	39.00	43.30	81.00	90.00
5	64-QAM	2/3	1	52.00	57.80	108.00	120.00
6	64-QAM	3/4	1	58.50	65.00	121.50	135.00
7	64-QAM	5/6	1	65.00	72.20	135.00	150.00
8	BPSK	1/2	2	13.00	14.40	27.00	30.00
9	QPSK	1/2	2	26.00	28.90	54.00	60.00
10	QPSK	3/4	2	39.00	43.30	81.00	90.00
11	16-QAM	1/2	2	52.00	57.80	108.00	120.00
12	16-QAM	3/4	2	78.00	86.70	162.00	180.00
13	64-QAM	2/3	2	104.00	115.60	216.00	240.00
14	64-QAM	3/4	2	117.00	130.00	243.00	270.00
15	64-QAM	5/6	2	130.00	144.40	270.00	300.00
16	BPSK	1/2	3	19.50	21.70	40.50	45.00
...	...	...	...	...	...	...	...
31	64-QAM	5/6	4	260.00	288.90	540.00	600.00

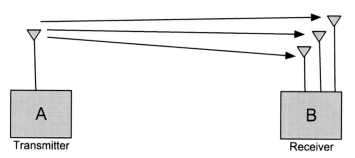

Fig. 2.6 An example 1 × 3 MIMO link

is much higher due to their constructive addition. Their results also demonstrate that MRC with only two antennas already shows large improvements, but MRC with three antennas shows an even further, though smaller, improvement.

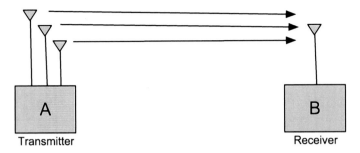

Fig. 2.7 A 3 × 1 MIMO link

2.2.1.2 Transmit Diversity

Similar to receiver diversity, transmit diversity techniques apply to cases where there are multiple antennas at the transmitting node and a single antenna at the receiving node, such as the 3 × 1 case in Fig. 2.7. There are two widely used methods of transmit diversity.

1. Transmit Beamforming: The technique can be considered an informed inverse of the MRC technique. In transmit beamforming, the transmitter precodes the signals sent from antennas such that their phase have an opportunity of constructive addition at the receiver antenna. As in MRC, the signals can be weighted using expected SNR of each independent spatial path. This technique requires prior knowledge of path quality, which in turn requires feedback from the receiver. 802.11n uses various control packets in order to notify the transmitter regarding the path statistics. Phased antenna arrays can also be used for beamforming in which phase delays are added via their physical orientation so that the resultant signals meet constructively at the receiver. Note that this is different from switched beamforming where out of many available antenna one or more are chosen at any given time in order to establish best spatial path.
2. Space-time Codes: The idea behind space-time codes is to achieve transmitter diversity by encoding information in both spatial and temporal domain. This is done by replicating the data stream, encoding it using space-time codes and sending them out over different antennas. The space-time codes (Goldsmith 2005; Oestges and Clerckx 2007; Tse and Viswanath 2005) ensure that they are orthogonal in terms of their mutual interference so that the receiver can construct a strong signal. Due to their simplicity, and no requirement of feedback, they are often adopted for 802.11n systems.

Both transmit and receive diversity techniques can be implemented, together, to yield advantages of both techniques. These techniques allow sending (or receiving) the same data stream across multiple antennas for an improved and robust communication. On the other hand, spatial-division multiplexing can be used to exchange independent data stream at each antenna pair.

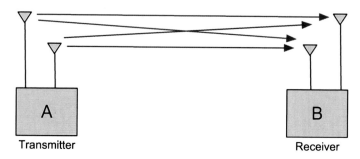

Fig. 2.8 Diversity techniques (receive and transmit) or spatial-division multiplexing can be used to yield greatest advantage of $N \times N$ MIMO system

Table 2.2 Theoretical achievable gains of using N antennas at end-points

	SISO	$1 \times N$ or $N \times 1$ diversity	$N \times N$ diversity	Multiplexing
Capacity	$B \log_2(1+\rho)$	$B \log_2(1+\rho N)$	$B \log_2(1+\rho N^2)$	$BN \log_2(1+\rho)$

2.2.2 Spatial-Division Multiplexing

Consider Fig. 2.8 in which there are N parallel stream between sender and receiver. These allows N independent spatial paths on which N different data streams can sent, and the receiver is able to receive these streams in parallel using dedicated RF chain processing. Foschini and Gans (1998), Halperin et al. (2010) outline the performance gains that can be achieved in systems with receiver diversity, transmit diversity and spatial-division multiplexing. These results are listed in Table 2.2.

In case of SISO systems, Shannon's theory gives us the capacity with B being the bandwidth of link. In a system with N antennas on receiver or transmitter side ($1 \times N$ or $N \times 1$ systems) the diversity techniques explained above can result into N times improvement in SNR. In the case of N antennas at each end, with diversity techniques implemented at both ends, a total of N^2 times increase of SNR can be achieved. In the case where spatial multiplexing is used to transmit N independent streams, the resultant benefit is N times the capacity that is achievable using a SISO system.

2.2.2.1 Experimental Evaluation of Throughput Gains of 802.11n

802.11n and inbuilt MIMO techniques have shown the potential of significant throughput increase when utilized in wireless mesh networks. Shrivastava et al. (2008) first presented a comprehensive experimental evaluation of 802.11n link by implementing them on a real testbed. They studied the impact of MIMO diversity, coexistence with other 802.11 networks, channel width and frame packet aggrega-

Fig. 2.9 Multiple packets can
be aggregated to generate an
Aggregate MAC Service Data
Unit

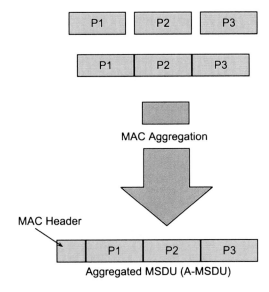

tion. 802.11n allows formation of Aggregate MAC Service Data Unit (A-MSDU) where multiple packets destined to a single destination are aggregated to create a large MAC frame (upto 7935 bytes). The process is illustrated in Fig. 2.9.

Shrivastava et al. (2008) experimented with one MIMO link with 3 × 3 settings in indoor environment, to observe the impact of channel width (20 or 40 MHz) and frame aggregation, for two different packet sizes (600 and 1200 bytes). As expected their results show that 40 MHz channels improve throughput over 20 MHz channels, as does aggregation. The throughput observed is larger for the larger packet size.

In practical terms, another important issue is the coexistence of 802.11g networks, and the effect on 802.11n links. The study by Shrivastava et al. (2008) shows, as expected, that a colocated 802.11g network adversely affects the throughput of the 802.11n network; significantly so, when the 802.11g link transmits at lower rates. The effect vanishes at higher transmission rates of 802.11g link; this is ascribed to the fact that at the higher rate, 802.11g uses the same modulation as 802.11n, hence is more compatible. Also, as before, the 40 MHz channel with aggregation performs well to combat the external interference. Apart from this, other cross-technology interference (baby monitors, cordless phones, microwave oven etc.) in ISM band has also been shown to reduce 802.11n throughput. This was initially identified by Bandspeed (2010), Cisco (2010), Miercom (2010), and was recently addressed by Gollakota et al. (2011). Some of the other solutions for the problem has been suggested by Cisco (2007), Lakshminarayanan et al. (2009), Moscibroda et al. (2008), Rahul et al. (2008).

2.3 Multihop Cellular Networks (MCN)

By using a WLAN technology for link layer communications, 802.11 based mesh networks explicitly leave the question of multi-hop paths to higher layer protocols, such as IP. This seems a natural development since 802.11, targeted at a local area span and context, has always depended on IP or other technology for wider area access. However, link layer technologies for cellular wireless networks, though conceptually also designed for single-hop communication with the base station, were targeted at wider areas of coverage. Thus it seems more natural to extend them for multi-hop paths within their own purview, and there have been advances along these lines in recent times.

In the last decade, cellular networks have leveraged a large number of physical layer technology such as CDMA, OFDMA etc. With other augmenting techniques like MIMO, they have become a strong contender for broadband wireless access networks. The most important advantage of cellular networks is the communication range of the cell tower, or base station. The larger communication range further allows better mobility management for highly mobile clients (e.g. a moving vehicle) as compared to 802.11 based systems. These advantages notwithstanding, cellular networks face various challenges. The first and foremost challenge is to meet the ever increasing traffic demand of clients. The data rates of cellular networks are typically lower than their 802.11 counterparts. A second issue is the design of cellular network to minimize the number of coverage holes. The users at the edge of the cells often face degraded services. A widely used solution to the problem is to use smaller cells which can well cover the desired area with sufficient quality of service. The downside of the solution is that this increases the cost of deployment dramatically.

Multihop cellular networks (MCN) (Oyman et al. 2007) use a different strategy to deal with the issues of performance and coverage. They deploy lightweight relay stations (RS) into cells that can relay the data between the base stations (BS) and mobile stations (MS). Several cellular network standards for 4G services have considered relaying in their drafts. As an example, WiMAX has included relaying in an amendment called IEEE 802.16j. Similarly, the recently released 3GPP Relase 10 Long Term Evolution-Advanced standard for IMT-advanced (4G) includes relaying stations. We next discuss both these MCN technologies from the aspect of their support of relaying.

2.3.1 IEEE 802.16: WiMAX

The IEEE 802.16 (Andrews et al. 2007; Eklund et al. 2002; Ghosh et al. 2005) working group was formed in 1999, and the first draft for point-to-multipoint, line-of-sight (LOS) communication with immobile users was proposed in 2004. This was later improved to accommodate non-LOS communication with mobile users in the draft standard of 2005. The draft has been widely known as 802.16e standard or mobile WiMAX, though officially it was merged into the 801.16-2009 standard, the

Air Interface for Fixed and Mobile Broadband Wireless Access System. To address the issues of performance and coverage, 802.16e was extended to incorporate multihop relaying. The task force derived a standard for 802.16 relays that is known as 802.16j, drafted to allow devices to provide backward compatibility with 802.16e (since 802.16e was standardized as early as 2006). 802.16j devices do not require any modifications to 802.16e based mobile devices, while the BS needs to be updated in order to accommodate relays.

2.3.1.1 Motivations for 802.16j

802.16j (referred to as 16j here onwards) was designed to address various design challenges. These challenges and the solutions by which 16j can address them are listed below.

1. Coverage: Relay stations can be used to solve the coverage problem in two ways. First, the locations where there exists a coverage hole due to significantly low signal strength from the BS can be now covered using RSs. An RS in such a case provides coverage to an area which is already within the ideal coverage region of BS (Fig. 2.10). The advantage of using RS instead of another BS is that typically RSs are cheaper and lightweight as compared to BSs. The solution works well especially in covering indoor coverage holes or other shadowed regions. Second, RSs can be used to extend the coverage of a BS in a specific region. Such regions are typically not within the coverage of BS but in near proximity (edge of BS) where deploying another BS is not cost-effective or otherwise not viable. This is shown in example in Fig. 2.11. This method also has wide application in terms

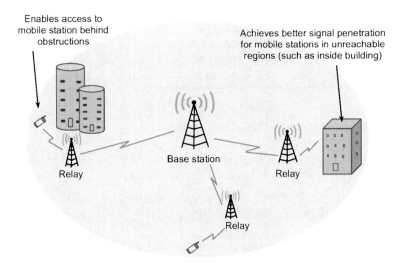

Fig. 2.10 Relay stations used at coverage holes in multihop cellular networks

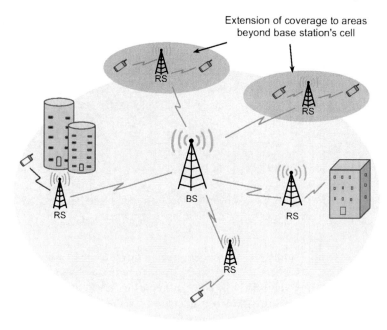

Fig. 2.11 Relay stations used for coverage expansion in multihop cellular networks

of coverage expansion. In coverage expansion, regions with no BS deployed, but closer to a coverage area, can be covered using an RS.

2. Performance: There may be regions in the coverage area of a BS that generate high traffic demand, which cannot be directly satisfied by the BS. Such clustered traffic demand places (parks, event venues etc.) can be further served using a RS. In such case, the purpose of deploying is to meet the localized traffic demand that can not be otherwise met by the BS. RSs can also be deployed in order to meet certain fast moving vehicles (such as trains, buses etc.) that have fixed routes and are expected to generate a large traffic demand. The low cost and ease of deployment make RS an appropriate choice for such cases (Fig. 2.12).

2.3.1.2 Relay Modes and Scheduling

The relays in 16j can be of two types: transparent and non-transparent. We define them below, identifying key differences between the modes.

- Transparent Mode: In transparent mode, framing and synchronization information is not forwarded by RSs but instead MSs receive the information from the BS. The main purpose of deploying RSs in such mode is to increase the capacity. The transparent RSs are within the coverage area of the BS and do not provide coverage extension because MSs are still dependent on the BS from framing and synchronization information. Transparent RSs are low complexity and their cost is lower than that of non-transparent (defined below). The scheduling of transmission

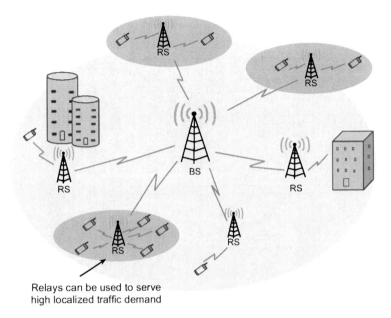

Relays can be used to serve
high localized traffic demand

Fig. 2.12 Relay stations used for localized high traffic demand and fast moving fixed-route vehicles in multihop cellular networks

between MS and RS is handled by the BS (called centralized scheduling). Every RS in transparent mode is connected directly to the BS, hence the maximum number of hops from the MS to the BS can not no more than 2.

- Non-transparent Mode: In non-transparent mode, RSs generate their own framing and synchronization information, and forward them to the MSs. The main purpose of deploying RSs in this mode is to expand the coverage. The capacity increase achieved by such RSs is not very high due to possible inter-RS interference. Their cost is typically higher than transparent RSs. They support distributed scheduling where RSs and their MSs coordinate in frame transmission. Non-transparent RSs can be interconnected to create topologies where number of hops between MS and BS can be more than two.

Note that since the original 802.16e standard was not designed to support relaying, 16j included certain modifications that can enable relay support while maintaining the backward compatibility with 16e devices. The modifications are mostly at MAC and PHY layers. We discuss these modifications next.

2.3.1.3 PHY Layer Enhancements

The original frame structure of 16e frames included two subparts—uplink (UL) and downlink (DL). These semantics made sense because the communication was always

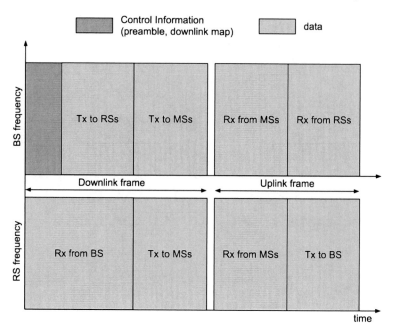

Fig. 2.13 Frame structure of 802.16j in transparent relay mode

between BS and MS. With the added support of RSs in 16j, it became necessary to support BS-RS and RS-MS communications, and stretch these semantics.

Transparent Relay Stations (T-RS) Frame Structure

As we remarked, in transparent relaying, frame and synchronization information is sent by the BS directly to the MSs. This is shown in Fig. 2.13. The DL frame in transparent mode is divided into two zones:

- Access zone: In the access zone of the DL frame, the BS first sends out information to RSs, as well as MSs directly connected to the BS. During this period, RSs receive from the BS.
- Transparent zone: In the transparent zone, the RSs transmit to their MSs while the BS can transmit to the MSs it is directly connected to.

The BS-RS and RS-MS communications that might happen at the same time during the transparent zone of the downlink period can be achieved by providing different frequencies for BS and RS transmissions. The uplink transmissions begin after the downlink period. As with downlink, uplink period is also divided into two zones:

- Access zone: During this period, the mobile stations receive from the BS or the RS.

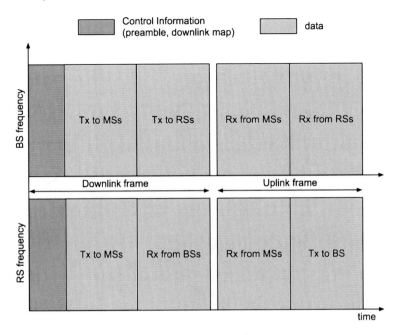

Fig. 2.14   Frame structure of 802.16j in non-transparent relay mode

• Relay zone: In the relay zone, RSs transmit their data to BS.

Since the maximum number of hops in transparent relaying is no more that two, the above mentioned division of UL and DL works well. The same can be defined when using non-transparent relaying as below.

Non-Transparent Relay Stations (NT-RS): Frame Structure

In non-transparent relays, framing and synchronization information is sent by the RSs, in addition to by the BS. This is shown in Fig. 2.14. In this case, during DL access zone, BS and RSs transmit information to their associated MSs. During the DL relay zone, BS sends out information to RSs. During the UL access zone, MSs send information to their BS or RS, while during UL relay zone, RSs transmit information to the BS.

Note that this is simple when there are only two hops in non-transparent topology, but the case where there are more than two hops between MS and BS require more attention. The problem can be solved by having multiple relay zones in UL and DL as shown in Fig. 2.15. The hierarchical handling of RSs and MSs requires the introduction of more zones, with some stations inactive in certain zones to let the information percolate through the hierarchy.

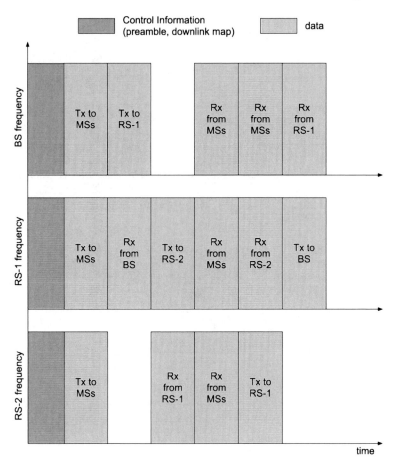

Fig. 2.15 Frame structure 802.16j in non-transparent relay mode with two levels of relay stations

2.3.1.4 MAC Layer Modifications: RMAC and Tunneling

As we noted before, link scheduling in MCNs can be centralized (transparent RS) or distributed (non-transparent RS). Also, it is worth noting that data exchange between MS and BS is in general connection-oriented. This means that every connection initiated by MS or BS receives a unique connection ID. In the case where a MS connects to a RS, the connection ID is provided by the RS. To further support connections, 16j includes a MAC protocol called R-MAC. In R-MAC, various connections initiated at MSs connected to a RS can be treated as a single connection from the point of view of other intermediate RSs. This way, tunneling abstracts the difference between various connections for the intermediate RSs. The access RS and the BS can interpret the tunneled connections. The tunneling support of R-MAC protocol has multiple advantages. First, it ensures that MSs are unaware of intermediate RSs in order to provide backward compatibility with 16e MS devices. Second, since one of the goals of relays is to satisfy highly localized traffic pattern, multiple MS connections from such a hot-spot can be treated as a logically stand-alone connection. During procedures like handoffs, this tunneled connection can be handed over to another cell as if all the MSs of the tunneled connections are moving together.

Other specific issues which are the topics of active research, such as relay placement (Lin et al. 2007), security (Dai and Xie 2010) etc. involved in 16j design are discussed in later chapters.

2.3.2 3GPP LTE-Advanced Relaying

The ongoing development of the Long Term Evolution standards by the 3GPP organization to meet the International Telecommunication Union's requirements of 4G cellular standards provides another example of the introduction of relaying into a framework originally designed for single- or two-hop communication. ITU (ITU-R 2008) has stated the following requirements for realizing true 4G mobile systems:

- High mobility environment (speed <350 Kms/h)

  - Peak data rate of 100 Mbps
  - Average case latency of 100 ms

- Low mobility environment (speed <10 Kms/h)

  - Peak data rate of 1 Gbps
  - Average case latency of 10 ms

Systems using the 3GPP LTE-Advanced (Abeta 2010; Bai et al. 2012; Ghosh et al. 2010; Lo and Niemegeers 2009; Mogensen et al. 2009; Sawahashi et al. 2009; Wirth et al. 2009; Yang et al. 2009) Release 10 (currently under process of standardization at ITU-T) have the potential to achieve these requirements. LTE-A includes advanced physical layer technologies such as carrier aggregation etc. and also includes relaying.

As in the case of 802.16j, the purpose of relaying in LTE-A systems is twofold, embodied by two types of relay stations proposed.

- Type-1 Relay Stations: They are similar to 802.16j non-transparent stations. Their purpose is to extend coverage to MSs beyond the coverage region. Conceptually the only differences between Type-1 relay stations of LTE-A and non-transparent relay stations of 802.16j are that LTE-A does not allow more than two hops in relaying, in order to guarantee improved latency.
- Type-2 Relay Stations: They are similar to 802.16j transparent stations. Their purpose is to improve the signal quality and quality of service to MSs within the cell of the BS.

Apart from the differences mentioned above, relaying in LTE-A and 802.16j standards are very similar in concept. Individual design problems of relaying in MCNs will be discussed further in later chapters.

2.4 Cognitive Radio Networks

In the last decade, the proliferation of wireless technology standards have given rise
to the problem of spectrum scarcity. This is due to the fact that spectrum allocation authorities have traditionally used fixed block assignment scheme for newer
technologies. As an example, such a problem has been reported by the US FCC
(FCC 2002). Depending on the current utilization of wireless technologies, it has
been observed that certain blocks of spectrum are underutilized while other parts are
overly congested. The 400–700 MHz spectrum block that is only utilized sporadi-
cally provides an example, while the ISM bands (especially 2.4 GHz) are excessively
crowded.

Dynamic Spectrum Access, or *cognitive radio* technology can be used to mitigate
the spectrum scarcity. Cognitive radios can dynamically access the spectrum when
it is not in use. The term was first introduced by Mitola and Maguire (1999) and
subsequently used by seminal work such as by Akyildiz et al (2006), Haykin (2005).
A cognitive radio has the ability to sense the medium widely, re-configure itself to
transmit in some targeted spectrum, and thus utilize the medium dynamically. An
unutilized spectrum block (typically known as "white space") can be exploited in
temporal, spatial and frequency domain in order to use it for communication. In the
context of cognitive radio, spectrum users can be divided into two classes—primary
and secondary users. The primary users are incumbent users who have licensed
access to the spectrum block, and their access to the block must be given the highest
priority. On the other hand, the secondary users access the spectrum opportunistically
whenever the primary users are not using the spectrum. This is shown in Fig. 2.16.

2.4.1 Cognitive Mesh Networks

The cognitive radio technology holds a special importance in design and development
of wireless access networks especially wireless mesh networks. This is because one
of the most widely adopted wireless standard—802.11 standard operates in the ISM
band. The current infrastructure deployments of 802.11 has resulted in congestion
in the ISM band (Akella et al. 2005). Apart from this, other technologies such as
Bluetooth has resulted into the ISM band being excessively utilized. Since most of
the wireless mesh networks are deployed using 802.11 radio technology, they are
expected to further contend for access to the ISM band. To address the issue, cognitive
radios are necessary at each mesh node to detect and opportunistically switch to non-
congested channels. This can yield improved performance because of its dynamic
access to medium.

There are numerous design challenges when designing a mesh network where
mesh nodes opportunistically switch to vacant white spaces in order to improve the
performance. First and foremost, due to dynamic spectrum access, mesh nodes no
longer share a common control channel that can be used to exchange necessary
control information. It was shown by Zhao et al. (2005) that neighboring nodes

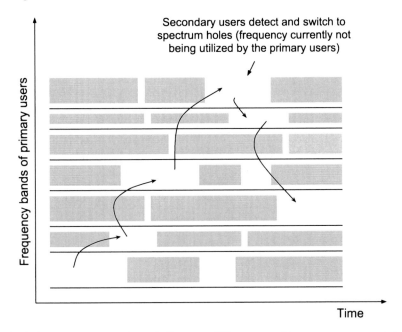

Fig. 2.16 Secondary users can dynamically access different parts of the spectrum opportunistically when they are not being utilized by the primary users

may have some common channels that are vacant for them simultaneously but the network-wide availability of a common vacant channel is very rare. This requires that nodes operate using a distributed control plane which in turn imposes numerous design challenges for upper-layer protocols.

In order to design efficient upper-layer protocols, it is first necessary to understand the interference relationship between primary and secondary users. There are two types of interference models largely used.

- Binary interference model: whenever there is any activity of primary users in a given channel, the channel becomes useless for secondary users.
- Interference temperature model: secondary users can communicate via a channel that is currently being utilized by primary users if the interference caused by secondary users to the primary users is below a certain pre-defined interference temperature threshold.

It is clear that interference temperature model is more general but further complicates the design problems.

2.4.2 IEEE 802.22

The issue of spectrum scarcity and under-utilization, and potential solution using cognitive radio has attracted tremendous interest from both research community and standardization bodies. One of the first standards to be developed using this cognitive radio technology is IEEE 802.22. The standard aims to utilize the unused spectrum of broadcast television service to provide broadband access to rural areas with low population density. These unused TV spectrum bands are often referred as TV white spaces. Even though the standard does not specify implicit support for multi-hop networking, in such cases mesh networking can be especially useful in rural regional networks.

2.4.3 TV White Spaces

The reports FCC (2004, 2006) outline how and which TV channels can be used for the purpose of rural broadband development. Figure 2.17 shows the spectrum and its channels that are made open as TV white spaces. As shown, channels above 700 MHz were auctioned to wireless service providers by the FCC in 2008. Due to the transition to digital television, FCC was able to free the TV white space block in 2009. These channels are 5–13 in the VHF band and 14–51 in the UHF band. The usage of the channel for secondary users is only permitted so that no interference is caused to the licensed TV subscribers and other low power devices such as wireless microphones.

Secondary users can either attempt to predict the activities of primary users or can use readily available information from any third party. In 802.22, there are two ways by which secondary users can perceive the activities of primary users.

- Geo-location database: In this method, devices equipped with GPS can query the central database using their location to determine the activity of primary users. This approach is especially useful for low-mobility or fixed devices.
- Spectrum sensing: Secondary users can sense the medium for its availability and utilize the information to make the transmission decision. This method is especially attractive since it does not require any central authority for decision making. On the other hand, the method is also very difficult to implement since even neighboring nodes might end up determining different information about the spectrum. This distributed sensing and decision making has attracted a lot of research which we will cover in later chapters.

Further details of 802.22 standard, its PHY and MAC layer considerations can be found in Stevenson et al. (2009). Figure 2.18 shows a network in which nodes of mesh network operate as secondary nodes to primary network of TV broadcast stations and its subscribers. Such networks have been studied by Akyildiz et al. (2009), Chen et al. (2008), Chowdhury and Akyildiz (2008) and others. In such a case, mesh nodes can have multiple frequency-agile radios to serve the associated clients and facilitate intra-mesh communications.

TV Channel No.	Start - end Frequency (MHz)
51	692 - 698
50	686 - 692
49	680 - 686
48	674 - 680
47	668 - 674
46	662 - 668
45	656 - 662
44	650 - 656
43	644 - 650
42	638 - 644
41	632 - 638
40	626 - 632
39	620 - 626
38	614 - 620
37	608 - 614
36	602 - 608
35	596 - 602
34	590 - 596
33	584 - 590
32	578 - 584
31	572 - 578
30	566 - 572
29	560 - 566
28	554 - 560
27	548 - 554
26	542 - 548
25	536 - 542
24	530 - 536
23	524 - 530
22	518 - 524
21	512 - 518
20	506 - 512
19	500 - 506
18	494 - 500
17	488 - 494
16	482 - 488
15	476 - 482
14	470 - 476
13	210 - 216
12	204 - 210
11	198 - 204
10	192 - 198
9	186 - 192
8	180 - 186
7	174 - 180
Channels used for FM, Ham and military applications	88 - 173
6	82 - 86
5	76 - 82
4	66 - 72
3	60 - 66
2	54 - 60

Ultra High Frequency (UHF) Band

Very High Frequency (VHF) Band

Fig. 2.17 TV white space spectrum

Fig. 2.18 Mesh routers operating as secondary users in holes of TV white space spectrum

One major advantage of TV white space is that FCC has not enforced any specific physical layer mechanisms (such as modulation etc.). This can allow the TV white spaces to be treated as an ISM band, and numerous devices, technologies and applications can be developed. (It should be noted, however, that the concern that white space networking may impact incumbents has led the FCC to mandate a tighter spectrum mask for TV white space use, so that existing 802.11 devices cannot be simply frequency-shifted and used in this spectrum.) In order to understand the success of such white space-based technology, it is first necessary to understand when and how much vacancy is indeed available in these channels. To this end, Chowdhury et al. (2011) first studied the availability of TV white spaces using USRP2 (Ettus 2009) radios. They observed a large variation in the mean received power on channels 21–51, indicating the potential for white space usage. They noted that the temporal behavior of the signal introduces further complexity.

While white space networks may be promising for the future of mesh networking, at this time they are far from being as mature as the existing technology we have previously described in this chapter. Especially from the point of view of designing mesh networks for predictable performance and behavior, cognitive radio technology appears to be a research horizon rather than a development one. Individual design problems such as sensing, collaboration among cognitive radio nodes, and their upper layer protocols will be discussed in later chapters.

References

Abeta S (2010) Toward lte commercial launch and future plan for lte enhancements (lte-advanced). In: IEEE international conference on communication systems (ICCS), pp 146–150. doi:10.1109/ICCS.2010.5686367

Akella A, Judd G, Seshan S, Steenkiste P (2005) Self-management in chaotic wireless deployments. In: Proceedings of the 11th annual international conference on mobile computing and networking (MobiCom '05). ACM, New York, NY, USA, pp 185–199. doi:10.1145/1080829.1080849

Akyildiz IF, Lee WY, Vuran MC, Mohanty S (2006) Next generation/dynamic spectrum access/cognitive radio wireless networks: a survey. Comput Netw 50(13):2127–2159. doi:10.1016/j.comnet.2006.05.001, http://www.sciencedirect.com/science/article/pii/S1389128606001009

Akyildiz IF, Lee WY, Chowdhury KR (2009) Crahns: cognitive radio ad hoc networks. Ad Hoc Netw 7(5):810–836. doi:10.1016/j.adhoc.2009.01.001, http://www.sciencedirect.com/science/article/pii/S157087050900002X

Andrews JG, Ghosh A, Muhamed R (2007) Fundamentals of WiMAX: understanding broadband wireless networking (Prentice Hall communications engineering and emerging technologies series). Prentice Hall PTR, Upper Saddle River

Bai D, Park C, Lee J, Nguyen H, Singh J, Gupta A, Pi Z, Kim T, Lim C, Kim MG, Kang I (2012) Lte-advanced modem design: challenges and perspectives. IEEE Commun Mag 50(2):178–186. doi:10.1109/MCOM.2012.6146497

Bandspeed (2010) Understanding the effects of radio frequency (RF) interference on WLAN performance and security

Chen T, Zhang H, Matinmikko M, Katz M (2008) Cogmesh: cognitive wireless mesh networks. In: IEEE GLOBECOM workshops, pp 1–6. doi:10.1109/GLOCOMW.2008.ECP.37

Chowdhury K, Akyildiz I (2008) Cognitive wireless mesh networks with dynamic spectrum access. IEEE J Sel Areas Commun 26(1):168–181. doi:10.1109/JSAC.2008.080115

Chowdhury K, Doost-Mohammady R, Meleis W, Di Felice M, Bononi L (2011) Cooperation and communication in cognitive radio networks based on tv spectrum experiments. In: IEEE international symposium on a world of wireless, mobile and multimedia networks (WoWMoM), pp 1–9. doi:10.1109/WoWMoM.2011.5986378

Cisco (2007) Cisco CleanAir technology. www.cisco.com/en/US/netsol/ns1070/index.html

Cisco (2010) Wireless RF interference customer survey result. White paper, c11–609300-00

Dai X, Xie X (2010) Analysis and research of security mechanism in ieee 802.16j. In: International conference on anti-counterfeiting security and identification in communication (ASID), pp 33–36. doi:10.1109/ICASID.2010.5551846

Eklund C, Marks R, Stanwood K, Wang S (2002) IEEE standard 802.16: a technical overview of the wirelessmantm air interface for broadband wireless access. IEEE Commun Mag 40(6):98–107. doi:10.1109/MCOM.2002.1007415

Ettus (2009) Universal software radio peripheral. www.ettus.com

FCC (2002) Federal communications commission spectrum policy task force, Report ET docket no. 02–135

FCC (2004) Et docket 04–186: Notice of proposed rule making, in the matter of unlicensed operation in the tv broadcast bands

FCC (2006) Et docket 08–260: Second report and order and memorandum opinion and order, in the matter of unlicensed operation in the tv broadcast bands additional spectrum for unlicensed devices below 900 MHZ and in the 3 GHz band

Foschini G, Gans M (1998) On limits of wireless communications in a fading environment when using multiple antennas. Wirel Pers Commun 6:311–335. doi:10.1023/A:1008889222784

Ghosh A, Wolter D, Andrews J, Chen R (2005) Broadband wireless access with wimax/802.16: current performance benchmarks and future potential. IEEE Commun Mag 43(2):129–136. doi:10.1109/MCOM.2005.1391513

Ghosh A, Ratasuk R, Mondal B, Mangalvedhe N, Thomas T (2010) Lte-advanced: next-generation wireless broadband technology [invited paper]. IEEE Wirel Commun 17(3):10–22. doi:10.1109/MWC.2010.5490974

Goldsmith A (2005) Wireless communications. Cambridge University Press, Cambridge

Gollakota S, Adib F, Katabi D, Seshan S (2011) Clearing the RF smog: making 802.11n robust to cross-technology interference. In: Proceedings of the ACM SIGCOMM 2011 conference on SIGCOMM (SIGCOMM '11). ACM, New York, NY, USA, pp 170–181. doi:10.1145/2018436.2018456

Halperin D, Hu W, Sheth A, Wetherall D (2010) 802.11 with multiple antennas for dummies. SIGCOMM Comput Commun Rev 40:19–25. doi:10.1145/1672308.1672313

Haykin S (2005) Cognitive radio: brain-empowered wireless communications. IEEE J Sel Areas Commun 23(2):201–220. doi:10.1109/JSAC.2004.839380

ITU-R (2008) Report m.2134: Requirements related to technical performance for imt-advanced radio interface

Judd G, Wang X, Steenkiste P (2008) Efficient channel-aware rate adaptation in dynamic environments. In: Proceedings of the 6th international conference on mobile systems, applications, and services (MobiSys '08), ACM, New York, NY, USA, pp 118–131. doi:10.1145/1378600.1378615

Lakshminarayanan K, Sapra S, Seshan S, Steenkiste P (2009) RFDump: an architecture for monitoring the wireless ether. In: Proceedings of the 5th international conference on emerging networking experiments and technologies (CoNEXT '09), ACM, New York, NY, USA, pp 253–264. doi:10.1145/1658939.1658968

Lin B, Ho PH, Xie LL, Shen X (2007) Optimal relay station placement in ieee 802.16j networks. In: Proceedings of the 2007 international conference on Wireless communications and mobile computing (IWCMC '07). ACM, New York, NY, USA, pp 25–30. doi:10.1145/1280940.1280947

Lo A, Niemegeers I (2009) Multi-hop relay architectures for 3gpp lte-advanced. In: 2009 IEEE 9th Malaysia international conference on communications (MICC), pp 123–127. doi:10.1109/MICC.2009.5431478

Miercom (2010) Cisco cleanair competitive testing

Mitola IJ, Maguire JGQ (1999) Cognitive radio: making software radios more personal. IEEE Pers Commun 6(4):13–18. doi:10.1109/98.788210

Mogensen P, Koivisto T, Pedersen K, Kovacs I, Raaf B, Pajukoski K, Rinne M (2009) Lte-advanced: the path towards gigabit/s in wireless mobile communications. In: 1st international conference on wireless communication, vehicular technology, information theory and aerospace electronic systems technology (Wireless VITAE 2009), pp 147–151. doi:10.1109/WIRELESSVITAE.2009.5172440

Moscibroda T, Chandra R, Wu Y, Sengupta S, Bahl P, Yuan Y (2008) Load-aware spectrum distribution in wireless lans. In: IEEE international conference on network protocols (ICNP 2008), pp 137–146: doi:10.1109/ICNP.2008.4697032

Oestges C, Clerckx B (2007) MIMO wireless communications: from real-world propagation to space-time code design. Academic Press, Oxford

Oyman O, Laneman N, Sandhu S (2007) Multihop relaying for broadband wireless mesh networks: from theory to practice. IEEE Commun Mag 45(11):116–122. doi:10.1109/MCOM.2007.4378330

Rahul H, Kushman N, Katabi D, Sodini C, Edalat F (2008) Learning to share: narrowband-friendly wideband networks. In: Proceedings of the ACM SIGCOMM 2008 conference on data communication (SIGCOMM '08). ACM, New York, NY, USA, pp 147–158. doi:10.1145/1402958.1402976

Sawahashi M, Kishiyama Y, Taoka H, Tanno M, Nakamura T (2009) Broadband radio access: Lte and lte-advanced. In: International symposium on intelligent signal processing and communication systems (ISPACS 2009), pp 224–227. doi:10.1109/ISPACS.2009.5383862

Shrivastava V, Rayanchu S, Yoonj J, Banerjee S (2008) 802.11n under the microscope. In: Proceedings of the 8th ACM SIGCOMM conference on Internet measurement (IMC '08). ACM, New York, NY, USA, pp 105–110, doi:10.1145/1452520.1452533

Stevenson C, Chouinard G, Lei Z, Hu W, Shellhammer S, Caldwell W (2009) IEEE 802.22: the first cognitive radio wireless regional area network standard. IEEE Commun Mag 47(1):130–138. doi:10.1109/MCOM.2009.4752688

Telatar E (1999) Capacity of multi-antenna gaussian channels. Eur Trans Telecommun 10(6):585–595. doi:10.1002/ett.4460100604

Tse D, Viswanath P (2005) Fundamentals of wireless communications. Cambridge University Press, Cambridge

Wirth T, Venkatkumar V, Haustein T, Schulz E, Halfmann R (2009) Lte-advanced relaying for outdoor range extension. In: IEEE 70th vehicular technology conference fall (VTC 2009-Fall), pp 1–4. doi:10.1109/VETECF.2009.5378969

Yang Y, Hu H, Xu J, Mao G (2009) Relay technologies for wimax and lte-advanced mobile systems. IEEE Commun Mag 47(10):100–105. doi:10.1109/MCOM.2009.5273815

Zhao J, Zheng H, Yang GH (2005) Distributed coordination in dynamic spectrum allocation networks. In: First IEEE international symposium on new frontiers in dynamic spectrum access networks (DySPAN 2005), pp 259–268. doi:10.1109/DYSPAN.2005.1542642

Chapter 3
Mesh Design: Lower Layer Issues

3.1 WMN Design Challenges

As with any kind of networks, research challenges in WMN design can be traced back to their characteristics and motivations of deployment. The reason why WMNs are often seen as the last few miles network is the potential of easy retro-fit: the coverage area of standards like WLAN can be extended further without the requirement of any specific infrastructure. There are numerous advantages of this over a traditional WLAN deployment. First and foremost, a WMN deployment can be modified or extended with ease depending on the coverage requirements. Since the mesh nodes do not require a wired connection, more mesh nodes can be added or existing mesh nodes can be moved to desired locations depending on current coverage requirement of the network. This also reduces the impact of error in initial RF measurements at the time of site survey, because the network can always be adjusted to changing coverage and performance needs. Second, with the use of renewable energy sources such as solar energy, it is in fact possible to achieve a completely tether-less deployment of mesh nodes. These advantages make WMN an especially attractive alternative for wireless access networks.

Even though WMN are expected to deliver high-performance wireless services, inherently they are ad-hoc in nature, which provides the added benefits of robustness and self-management. These imply a more ad-hoc model than the traditional infrastructure model of access networks. Such a change poses various challenges for designers. Maintaining scalability with expansion of the WMN, providing novel medium access and spatial reuse techniques, interference mitigation, and supporting heterogeneity among standards are a few of these challenges.

Every transmission between wireless mesh routers creates interference in its neighborhood, which is a major issue challenging the performance of WMNs. On one hand, a certain power level for transmission is necessary for successful reception at the receiver. On the other hand, high power transmissions cause high interference and MAC layer collisions at other unintended receivers. Various attempts have been made to model the effects of interference using abstract theoretical models as well as

P. H. Pathak and R. Dutta, *Designing for Network and Service Continuity in Wireless Mesh Networks*, Signals and Communication Technology, DOI: 10.1007/978-1-4614-4627-9_3,
© Springer Science+Business Media New York 2013

measurement-based models. Using the knowledge of interference from such models, researchers have designed protocols for power control, link scheduling, routing and channel/radio assignment. Energy conservation not being a typical objective, power control and topology control mechanisms in WMNs mainly deal with assigning transmission power levels to nodes such that the traffic demands are satisfied with better overall throughput. The concurrent objective of any such mechanism is also to reduce interference, which in turn increases the achievable network capacity.

Power control and *topology control* mechanisms determine the network connectivity and underlying physical layer topology. All links of such a topology can carry the traffic between the nodes, and the reception rate depends on the quality of the link. *Routing strategy* determines reliable and high throughput end-to-end paths between the source and destination of data. Various characteristics of links such as quality, stability and reliability play an important role in routing metric design which is used by the routing protocol. *Link scheduling* strategies estimate transmission conflicts between links of these routing paths using the interference model, and try to achieve a conflict-free feasible transmission schedule. There are various challenges in distributed implementation of any such scheduling scheme which combines medium access, collision detection/avoidance and transmission scheduling techniques. Spatial reuse (concurrent transmissions on more than one links) can be increased when non-interfering links are scheduled in parallel using intelligent scheduling. To further mitigate the interference effects, interfering links are often separated in the frequency domain. *Channel/radio assignment* schemes try to arrange nearby transmissions on orthogonal or minimally overlapping channels in single or multi-radio WMNs.

Clearly, many of these problems are coupled, and it is attractive to consider them in conjunction. When transmission power level of nodes change, the scheduling decision should be revised which may in turn require reallocation of power levels for certain nodes. Similarly, when channel assignment is performed, newer routing decisions should be made to accommodate the changes in connectivity; conversely, routing itself can help to make more intelligent decision about channel assignment. Researchers have increasingly focused on such problems of joint design in recent years, and we will discuss such problems in the next chapter.

Such coupled problems, while more attractive, are also more complex to formulate and solve. The success of such approaches is at least partly due to the large body of work previously accomplished on addressing these problems in isolation. In this chapter, we describe such work on WMN design from the literature. These approaches continue to be extremely necessary background, and provide valuable building blocks, when addressing problems of design with continuity issues in mind.

These areas have been previously surveyed in literature. The well-known survey of Akyildiz et al. (2005) focuses on the operations and problems on a layer by layer basis. Similarly, Nandiraju et al. (2007) and Bruno et al. (2005) survey design problems separately at each layer, and provide useful insights regarding standard specific deployment issues, respectively. Some of the relevant surveys are dedicated to specific design problem like multiple access protocols (Kumar et al. 2006a), specific techniques of improving spatial reuse (Alawieh et al. 2009), energy efficiency (Guo and Yang 2007; Jones et al. 2001), secure routing (Hu and Perrig 2004), multicast routing (Junhai et al. 2009), dynamic spectrum access (Akyildiz et al. 2006; Yucek and Arslan 2009), admission control (Hanzo and Tafazolli 2009), power control in sensor networks (Pantazis and Vergados 2007) etc. In this chapter, we focus on those problems that provide useful building blocks of survivability oriented design. Instead of surveying protocols developed at each layer, we focus on the fundamental problems and the operations like power control, link scheduling, routing etc. This approach is suitable for us, since many of the problems and protocols deal with functionalities related to more than a single layer. However, in the interest of logical grouping, we have organized our discussion into issues closer to the physical medium (this chapter), and network-wide issues (next chapter). This also aligns with our discussion of joint design issues and cross layering in a following chapter, since many of those approaches can be seen as an attempt to perform more than one of the functions *jointly*; for example, to provide an algorithm that performs both routing and link scheduling, so as to enable a more efficient link scheduling by choosing not to use high-interference links for certain routing paths.

Table 3.1 summarizes this categorization of the research we address in this chapter, and identifies the research problems that will later be seen to be building blocks for joint design research. In this table, we have cited a few contributions in the literature for each area. These are meant to be representative of the area, and to provide one of several alternate starting points to reading the literature of that area. We certainly do not imply any value judgement, nor do we imply this to be sufficient introduction with the field; they are starting points and not completion points. In the rest of this chapter and its bibliography we cite other important literature from each area that we are aware of.

3.2 Measuring and Modeling the Effects of Interference

One fundamental requirement for designing any WMN protocol is tractable yet realistic consideration of interference. The nature and impact of interference is highly unpredictable, which challenges the design of all upper-layer protocols. Researchers have proposed various ways to model the impact of interference; some salient ones are discussed below.

1. *Protocol Interference Model* (Gupta and Kumar 2000):
   Communication between nodes u and v results in collision-free data reception at node v if no other node within a certain interference range of v is transmitting simultaneously. This model has been further extended to consider link layer reliability using acknowledgments in which interference with either the forward transmission from u to v, or with the acknowledgement transmission from v to u, would hinder the data transfer. Thus any other transmission within the reception range of node u is also counted for interference. This is often referred as the *disk model* (or *double disk model*, for the case with acknowledgements) where interference is assumed to be a binary phenomena developed within a certain

Table 3.1 Categorization of WMN problems (lower layers)

Interference measurement and modeling (Sect. 3.2)—Tractable yet realistic estimation of interference in dynamic wireless environment **Objective:** Design of abstract interference models to aid upper layer protocol design and their comparison to actual measurements, link and network capacity analysis **Sample Literature:** Protocol and physical interference models (Gupta and Kumar 2000), scalable measurement based estimation of interference and packet delivery (Reis et al. 2006)
Power control (Sect. 3.3)—Assigning transmission power levels to nodes having transmission requirements **Objective:** Minimizing interference, avoiding MAC collisions for better network capacity and throughput, energy conservation (some special cases of WMNs) **Sample Literature:** Motivations and requirements of power control mechanism (Kawadia and Kumar 2005), uniform power assignment (Narayanaswamy et al. 2002), variable range power control (Li et al. 2003a)
Topology control (Sect. 3.4)—Choosing or avoiding certain links in network **Objective:** Interference mitigation and reducing MAC layer collisions **Sample Literature:** MST-based low interference topology design (Burkhart et al. 2004)
Link Scheduling (Sect. 3.5)—Scheduling link transmissions to achieve feasible and conflict-free transmission schedule **Objective:** Higher throughput and better spatial reuse, efficient medium access and utilization, fairness **Sample Literature:** Stability property for scheduling in multi-hop networks (Tassiulas and Ephremides 1992), link scheduling in protocol interference model (Jain et al. 2003) and physical interference model (Brar et al. 2006)
Channel/radio assignment (Sect. 3.6)—Assigning multiple channels to single or multiple radios at nodes **Objective:** Separation in frequency domain to increase concurrent transmissions and thus throughput **Sample Literature:** Motivations and challenges in multi-channel multi-radio mesh (Kyasanur et al. 2006), channel assignment using interference conflict graph based vertex or edge coloring (Subramanian et al. 2007) multi-radio conflict graph based centralized channel assignment (Ramachandran et al. 2006)

fixed distance from the source and the destination of any active link (shown in Fig. 3.1a: the top transmission interferes with correct reception of the one in the middle). Such Interference range of any node is often assumed to be larger than its communication range, by a constant factor such as 2.

2. *Physical Interference Model* (Gupta and Kumar 2000):

   Communication between nodes u and v results in collision-free data reception at node v if SINR (Signal to Interference and Noise Ratio) at node v is above a certain threshold β. If P_{vu} is the signal power received at node v from u, a packet from node u is successfully received at node v *iff*:

$$\frac{P_{vu}}{N + \Sigma_{i \in I} P_{vi}} \geq \beta \tag{3.1}$$

where I is a set of nodes simultaneously transmitting, N is the background noise and β is a physical layer dependent constant. β is typically referred as the receiver

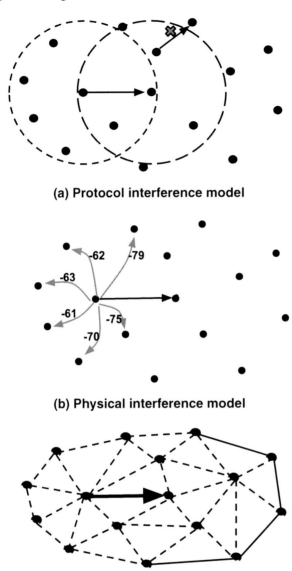

(a) Protocol interference model

(b) Physical interference model

(c) K-hop interference model

Fig. 3.1 Simple models of interference between transmissions: **a** the transmission marked with a × interferes with the other one, **b** the transmission contributes to interference at a number of other receivers, **c** only the solid links on the periphery, out of 2-hop range of the transmission, are free from interference

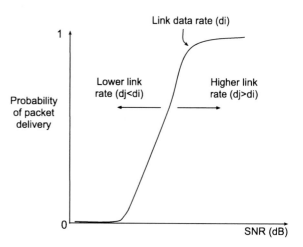

sensitivity, and its value can in turn determine the rate at which the link operates
between the node u and v. This is shown in Fig. 3.2. When SINR value increases
to a threshold determined by receiver sensitivity, the packet delivery ratio sharply
increases. Attempting to increase the link data rate with the same receiver sensitivity results in a higher value of this threshold SINR—the entire characteristic
curve shifts to the right. For a higher value of receiver sensitivity, a higher link
data rate is also observed.

The threshold based version of this SINR model was extended to a more general graded probabilistic SINR model (Santi et al. 2009) which also considers
SINR lesser than the threshold and predicts the probability of successful reception. This is shown in Fig. 3.3. By generating thousands of samples of received
packets on a link, Santi et al. (2009) showed that the relationship between percentage packets delivered to their corresponding SINR is not strictly binary, leading
them to formulate a probabilistic graded interference model that can capture this
relationship.

3. *K-hop Interference Model* (Sharma et al. 2006):
   In this model, no two links within K hops from each other can successfully
   transmit at the same time. The simplest case of such a model (with $K = 1$) is
   often referred as the node-exclusive interference model, where the only restriction
   imposed by interference is that a node can not transmit and receive on two separate
   links concurrently.

The above mentioned interference models can be further generalized by representing the interference relationship of links using a *conflict graph*. In a conflict graph,
every link in the network is represented as a vertex and two vertices share an edge if
and only if the corresponding links interfere with each other. Depending on the interference model and its directionality characteristics, the resultant conflict graph can
be undirected (double-disk model or k-hop interference model) or directed (physical interference model). An example of this shown in Fig. 3.4, following Grönkvist
(1998).

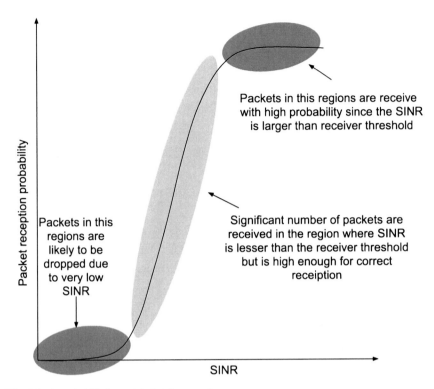

Packets in this regions are receive with high probability since the SINR is larger than receiver threshold

Packets in this regions are likely to be dropped due to very low SINR

Significant number of packets are received in the region where SINR is lesser than the receiver threshold but is high enough for correct receiption

Fig. 3.3   A probabilistic model of delivery ratio

3.2.1 Directional Antennas

All the above models assume that omni-directional antennas are used at mesh routers of WMNs. Recently, directional antennas have also been considered as a way of increasing the throughput capacity. Such antennas radiate energy asymmetrically (Figs. 3.5, 3.6), usually predominantly in one or a few directions ("beam-forming"), which enables a transmission to reach the desired destination, while causing less interference in the rest of the network. Though directional antennas improve the overall spatial reuse, they pose various other challenges in network design due to their directionality characteristics. As an example, inclusion of directional antennas require careful adaptation of the above mentioned interference models. In such a case, transmission by a node using directional antenna of beamwidth θ causes interference in a physical sector of angle θ with radius equals to its interference range (Yi et al. 2003; Zhang and Jia 2009).

Modeling link quality, capacity and the effect of interference can be an extremely difficult task as the wireless environment is often a complex combination of various parameters. Some such parameters and their interaction in an outdoor mesh environment was studied by Aguayo et al. (2004). With detailed experimentation, the study concluded that most of the lossy links in such environments are loosely correlated to

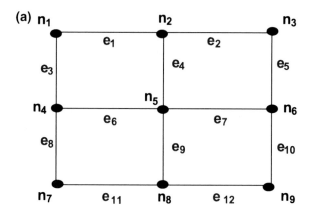

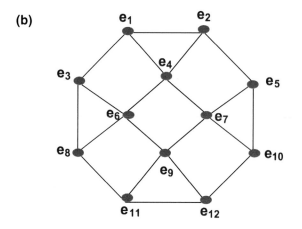

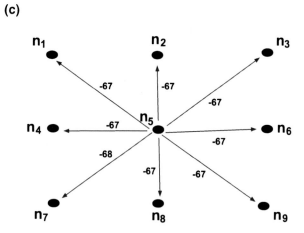

Fig. 3.4 The use of conflict graphs to represent interference: **a** network topology, **b** conflict graph using protocol interference model, **c** part of conflict graph created using physical interference model

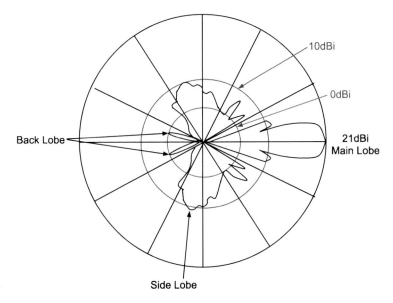

Fig. 3.5 Typical signal strength distribution for beamforming antenna

link distance and SNR values, but strongly related to multi-path fading of the environment. Such complex interaction of these parameters probably can not be modeled by the abstract interference models described above and requires some way to model more real-time dynamic wireless conditions. Some unrealistic assumptions made by abstract wireless interference models, and the consequent mismatch of simulation results from realistic conditions, were studied in Kotz et al. (2004) and Iyer et al. (2006). Some research (Shi et al. 2009) has tried to bridge the gap between protocol and physical interference models. It shows that it is in fact possible to preserve the advantage of the binary and geometric nature of the protocol model, if results that are produced using the protocol model are revisited with suitable methodology, and the corrected interference range is utilized in the simulations.

3.2.2 Utilizing Measurements for Modeling

Recently, researchers have attempted to rely on actual measurements to capture the effects of interference. If there is a way to feed the analytical models with realistic measurements like link quality, packet delivery rate etc., such models can accurately predict the interference effects. Completely depending on measurements to estimate interference might also raise a question of scalability since a large number of measurements can become intractable. Realizing the importance of measuring the interference, Padhye et al. (2005) presented an initial solution to the problem of scalability. An n node network may require $O(n^4)$ measurements for measuring the

Fig. 3.6 Interference relations are changed by the use of
beamforming antennas: **a** link
interference in nodes with
omni-directional antennas,
b link interference in nodes
with directional antennas

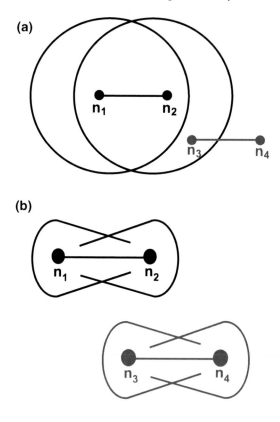

pair-wise interference of all possible set of wireless links. They proposed a notion of
Link Interference Ratio (LIR), which is a ratio of the total throughput of a set of links
when active together, versus their aggregate throughput when they are active individually. Comparing LIR values obtained on an actual 802.11 testbed, they observed
that most of the heuristics assumed in the literature for capturing effects of interference (including well used protocol interference model) are either too pessimistic or
too optimistic about their decisions of link interference and can lead to inefficient
upper layer protocols design. They go on to propose an empirical methodology to
approximate LIR values which requires only $O(n^2)$ actual measurements.

Such approaches, based on measurements, can nevertheless be time-consuming
and do not provide analytically tractable results when used under different network
settings. In contrast, Reis et al. (2006) proposes a measurement based model where n
measurements are seeded to a formulation (PHY model), which can then predict the
packet delivery rate and throughput with different sets of competing senders. This
PHY model modifies the traditional SINR model to use the actual measurements.
Such predictions can be then used with the MAC and traffic models for estimating
actual network performance. Similarly, Reis et al. (2006) and Kashyap et al. (2006)
also use measurements of RSSI (Received Signal Strength Indicator—measurement

of signal strength at the receiver's radio) and noise in commodity wireless cards, together with carrier sense factor values, to evaluate their effects on transmission capacity of nodes and delivery ratios of links.

With the same measurement based inputs, Qiu et al. (2007) extends the work of Reis et al. (2006) by modeling the interference and estimating the throughput among an arbitrary number of transmitting nodes (with unicast transmissions) and realistic traffic demands. First, with the consideration of 802.11 DCF and single-hop traffic, a generic N-node Markov chain model is presented where each state represents the set of nodes transmitting simultaneously. It is then extended for a sender model that estimates throughput, and a receiver model that estimates goodput; for saturated and unsaturated traffic demands, and in a broadcast transmission scenario. In the receiver model, slot-level loss rates of the Markov chain are converted to packet-level loss rates, which might be significantly higher—primarily due to the collisions with hidden terminals. Both the models are then extended for unicast transmissions, which capture retransmission and back-off in the sender model, and losses in the receiver model. Similarly, Kashyap et al. (2006) extends single interferer based PHY and MAC models for multiple interfering nodes and provides analytical solution for modeling link capacity in such a case.

As in the protocol interference model, interference is often assumed to be a deterministic on-off phenomenon for analytical tractability. In contrast to such a *binary* notion of interference, Hui et al. (2007) presents a Markov chain based model for *partial* interference to derive packet transmission and corruption probabilities. The study of Das et al. (2006b) studies *multi-way* interference, interference caused to a communication link by multiple transmitters. The authors show that even if a set of transmitters individually do not interfere with a given communication link, when they are active together they can cause significant interference to the link. This motivates moving beyond the LIR approximation of Padhye et al. (2005), which considers only two transmitters at a time, and points out the need of considering k-way interference possible from simultaneously active senders. Simulation and testbed experiments show that such multi-way interference is not wide-spread but can sometimes be significant. Improper estimation of interference can affect very basic functionality of the networking stack. Hamida et al. (2006) pointed out a previously unrevealed impact of interference in large scale multi-hop networks. They show that the set of discovered neighbors depends on the *frequency* of `hello` messages as well as the interference. The hybrid model they propose efficiently predicts the number of discovered neighbors and should be utilized to assist the `hello` protocol.

With several models and protocols depending on measurements, it is necessary that such measurements are accurate, and there exists an efficient way to collect them periodically without incurring much overhead. Kim and Shin (2006) presents a distributed approach for efficient measurements in which whenever it is possible, application traffic itself is used to probe the network, instead of specialized probing packets as commonly used in many measurement schemes. This results into lower overall probing overhead. While measuring the link losses, it is also important to distinguish between different types of causes for packet losses at various levels. The study of Ma et al. (2007) shows that packet losses can be classified into three types:

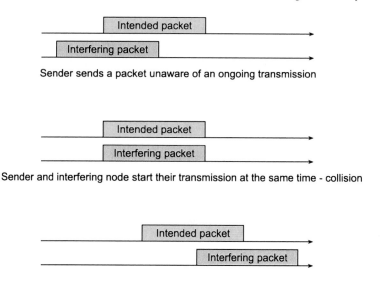

Fig. 3.7 Three types of losses caused by interference

Definition 3.1. Collision (Synchronous Interference): the intended packet and interfering packet start at the same time causing the intended packet to corrupted.

Definition 3.2. Asynchronous Interference—Type 1: signal strength prior of the interfering packet prior to the intended packet is strong enough such that the intended packet is dropped.

Definition 3.3. Asynchronous Interference—Type 2: signal strength of the interfering packets arriving subsequently after the intended packet can cause corruption of the intended packet.

Ma et al. (2007) also provide methodology to distinguish between the types of losses, and analyze their effects. Such a methodology is important for accurate interpretation of the relation between measurements and causes of packet losses (Fig. 3.7).

3.3 Power Control

3.3.1 Power Control and Topology Control

With many wireless networks, it is often not possible to perform intelligent node deployment due to geographical constraints. In such cases, the network topology depends on power control (PC) and topology control (TC) strategies for choice of links between the nodes. Such PC/TC decisions are crucial during optimization since all design decisions like link scheduling, channel assignment, routing are affected by the underlying network topology. The terms topology control and power control are often used interchangeably in literature since both attempt to control the transmission range of nodes while trying to achieve some desirable property of the topology. Considering the two control mechanisms from the global system-level perspective, *power control* strategies determine what power levels should be assigned to the nodes; the resultant topology is the supergraph from which a *topology control* mechanism can choose a subgraph that achieves a certain definite property like energy-efficiency, low interference etc. We make a distinction between the two; TC may be effected at layers higher than PC, by choosing not to make certain node adjacencies visible to the network layer (e.g. by filtering at the MAC layer). On the other hand, PC will almost invariably results in some effect on the topology, but the objective of PC may not be TC but the control of interference, or completely unrelated issues such as security, etc. We discuss approaches to TC that do not involve PC in the next section, and dedicate the rest of this section to PC approaches.

The problem of power control deals with assigning power levels to the nodes having transmission requirements in such a way that a particular objective is achieved, while still maintaining network connectivity as a fundamental requirement. Such an optimization objective can be lower interference, higher throughput and sometimes energy conservation. Power control mechanisms proposed in the literature can be broadly classified as follows.

3.3.2 Static Power Control

A static power allocation assigns power levels to the nodes that do not change frequently over time, unless there are drastic changes in the network topology. Such mechanisms are simpler and more robust, but often result in suboptimal performance due to their inefficient adaptation to changing traffic demands and dynamic wireless conditions. Static power control mechanisms presented in the literature can be further classified into uniform or non-uniform range power control.

3.3.2.1 Uniform Power Control

To understand the effect of transmission power on interference, refer to the example network shown in Figs. 3.8 and 3.9. Here, when the nodes have higher transmission power, transmission of each node interferes with a larger set of nodes. This in turn reduces the network capacity. On the other hand, when nodes use lower transmission power, it is in fact possible to maintain network connectivity while improving the network capacity. In seminal work on static power control, it was shown by Narayanaswamy et al. (2002) and Kawadia and Kumar (2005) that throughput is nearly optimal with all nodes operating at one common power level (COMPOW) that is the minimum required for maintaining network connectivity. It has been shown that such common power level can achieve the best case capacity $\Theta(1/\sqrt{n})$ bits/s (Gupta

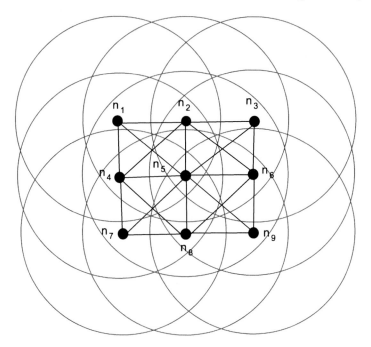

Fig. 3.8 Uniform power control—all nodes use the same power level for all transmissions

Fig. 3.9 COMPOW: the
common power level used
by all nodes is the minimum
necessary to keep the network
connected

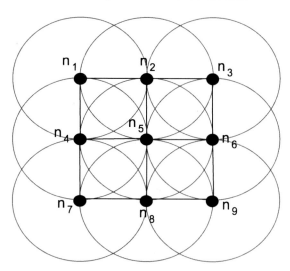

and Kumar 2000) with optimal node deployment. It is near optimal ($\Theta(1/\sqrt{n \log n})$
bits/sec) even in the case of random networks, without requiring any complex mechanisms.

The most important advantage of uniform power control is that it generally eliminates unidirectional links from the topology. We will see in later sections how unidirectional links can affect the network performance. The behavior of delay and throughput when changing transmission power control was studied by Kawadia and Kumar (2005). They showed that

> When (node-to-node) traffic load is high, a lower power level gives lower end-to-end delay, while under low load a higher power achieves lower delay.

Kawadia and Kumar (2005) show that at a certain power level, the delay increases slowly with increase in network throughput initially, but then switches to a higher rate of growth beyond a certain throughput. For higher power levels, the initial delay level is lower, but the switch to fast growing delay occurs sooner (with less throughput). This observation provides the above principle.

3.3.2.2 Non-uniform Power Control

Uniform range power control protocols like COMPOW have their disadvantages, one of which is that the common power level can be very high in non-uniform clustered topologies. The COMPOW protocol was extended for variable range power control by Kawadia and Kumar (2003). It describes three power control protocols—CLUSTERPOW, tunneled CLUSTERPOW and MINPOW. In CLUSTERPOW, the source node for any transmission uses a power level such that no other nodes in subsequent hops will need to use a higher power level. Figure 3.10 shows an example of the application of the CLUSTERPOW strategy, where the nodes use lower power levels (10 mW or even 1 mW) for intra-cluster links, while higher transmission power is used to establish inter-cluster links. This can be suboptimal and hence a recursive look-up mechanism is proposed in the tunneled version of the protocol. MINPOW uses the Bellman Ford algorithm with the power requirements as the cost function. Design of such protocols gave rise to many other variable range power control mechanisms, some of which we discuss next.

In variable range power control, different nodes in the network use different power levels. Such levels are often determined based on the node locations, overall network connectivity, tolerable interference and even routing paths. It was shown by Gomez and Campbell (2004) that variable-range transmission power control, where every node dynamically controls the transmission power, can outperform the COMPOW approach (Narayanaswamy et al. 2002) in terms of traffic carrying capacity of the network. It proposes a Minimum Spanning Tree(MST) based variable-range power control to maintain the network connectivity and increase the capacity. Specifically, it is shown that routing protocols based on such variable power levels can achieve twice the traffic carrying capacity than routing protocols based on common-range power control. An important result of Gomez and Campbell (2004) indicates that

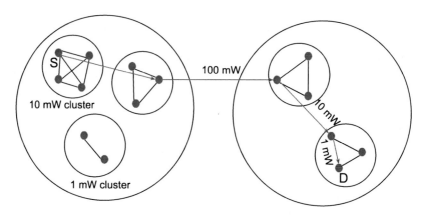

Fig. 3.10 CLUSTERPOW: using lower power intra-cluster than required inter-cluster

with variable power control, the average traffic carrying capacity remains constant even if more nodes are added to the network; this result is in contrast to the results presented by Gupta and Kumar (2000). Along the same lines, Li et al. (2003a) makes the first approach to overcome the disadvantages of uniform power assignment by building a decentralized Minimum Spanning Tree (MST) based topology; the algorithm presented uses transmission power or Euclidean distance (linear with power) as the weight of an edge in the network graph, and tries to build a minimum (power) spanning tree connecting all nodes. Every node determines its one hop on-tree nodes to be its actual neighbors, and the overall topology is built by integrating the MSTs of all nodes and maintaining symmetric links. Such a topology can maintain a lower node degree—which is shown to reduce interference and MAC-level collisions.

3.3.2.3 Non-uniform Power Control as a Design Tradeoff

There is an interesting trade-off between *longer-hops shorter-paths* and *shorter-hops longer-paths* data transfer in multi-hop wireless networks. It is shown by Behzad and Rubin (2005); Khalaf and Rubin (2004) that throughput and delay in 802.11 like networks can be optimized by using direct transmissions only. It claims that power control mechanisms should be based on per-link-minimality conditions, where nodes willing to transmit increase their power level just enough to reach the destination in a single hop. This is in sharp contrast to the results of Narayanaswamy et al. (2002) in which multi-hop routing paths are chosen between source and destination. This suggests that the fully (maximally) connected topology is always the optimum topology, independent of nodal distribution, traffic pattern and offered traffic load. They show that in a finite ad-hoc network, COMPOW does not yield maximum capacity. Under the assumption that all nodes have identical maximum power levels, they prove that the throughput capacity is maximized by all the nodes transmitting at their maximum required power. In spite of this trade-off, it is well-known that the COMPOW power level, which minimizes the overall interference level in the network, can achieve maximum *asymptotic* network capacity. In practical terms, the advantages of minimum or maximum power levels depend on several other factors like traffic pattern (node-to-node or node-to-gateway), network topology (uniform or clustered), etc.

When CSMA-CA based MAC is employed, increasing transmission power level of nodes results in more interference and collisions, which in turn reduces the throughput capacity. This is generally acknowledged due to extensive literature on power control (Kawadia and Kumar 2005) and topology control (Santi 2005), which shows that increasing transmission power of nodes results in larger collision domains; and due to random access nature of CSMA, overall spatial reuse reduces significantly when the transmissions are not separated in frequency or time domain. Pathak and Dutta (2011) showed that in TDM-scheduled WMNs, throughput behavior depends on *range-hop* tradeoff of various power control strategies. The trade-off is best characterized by two contradicting power control strategies (not specific to TDMA) proposed by *short-range-multi-hop* COMPOW (Narayanaswamy et al. 2002) and *long-range-single-hop* DirectTrans (Khalaf and Rubin 2004). COMPOW achieves better concurrency in link scheduling but requires more relaying at nodes with longer routing paths. In DirectTrans, nodes willing to transmit increase their power level until receiver can be reached in single hop; thus DirectTrans benefits from fewer transmissions and no relaying but suffers from lower spatial reuse due to long, high-interference links.

Figure 3.11 shows an example which demonstrates the trade-off. Assume that there are three flows $v_1 \rightarrow v_3$, $v_3 \rightarrow v_5$, $v_5 \rightarrow v_1$, each transferring one unit of traffic. Also, assume that shortest path routing is used and 2-hop interference model determines sets of mutually interfering links. In COMPOW topology (left), all six links have to be scheduled once to satisfy traffic demands of three flows. Links indicated by lines of the same style represent a set of links that do not mutually interfere. The routing path for each flow, in this case, contains two hops. When a TDM sched-

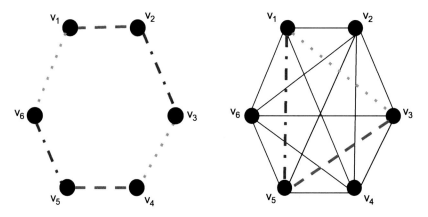

Fig. 3.11   The short-range-many-hops and long-range-single-hop tradeoff

uler is used to schedule the links, three slots are required to complete the transmission demands since two links can be scheduled in each of the three slot (e.g. v_1v_2 and v_4v_5). On the other hand, in DirectTrans topology (right), all nodes are directly in one hop reach of all other nodes (clique). In this case, only one transmission is required for every flow but each of the transmission interferes with all other transmissions (e.g. v_1v_3). The spatial reuse in this case drops to one, also requiring three slots for satisfying traffic demands. This way, even though COMPOW and DirectTrans are two extremes of uniform power control, their performance can become comparable in TDM-scheduled WMNs.

The network can be modeled using a unit disk graph $G_U = (V, E, \lambda)$ where V is the set of nodes and for any two nodes u and v, there exists an edge $uv \in E$ if their Euclidean distance $d_{uv} \leq \lambda$. COMPOW range (d_{min}) is defined as minimum value of λ such that G_U is connected. Similarly, DirectTrans range (d_{max}) is minimum value of λ such that G_U is *fully* connected (clique). We refer to the COMPOW graph of V as $G_C = (V, E_C, d_{min})$ and DirectTrans graph as $G_D = (V, E_D, d_{max})$. To understand the impact of changing power levels of nodes on throughput capacity in different topologies and traffic patterns, Pathak and Dutta (2011) uses uniform power *capability*. The behavior of the network is studied under different power levels; at any step, all nodes operate at the same power level. However, this does not imply uniform power control—the common level is a capability rather than a fixed transmission power. The increase in power levels can actually be interpreted as bounding the maximum transmission power of nodes. That is, if a node has an available high power level, this does not imply that it will always transmit at this increased power level. If a neighbor is reachable at a lower power level, transmissions to that neighbor will utilize that lower level.

Assuming the path-loss model of signal propagation, if transmitted signal power is P_t and distance between the transmitter and the receiver is d then received signal power (P_r) attenuates as $P_r \propto P_t(d^{-\alpha})$, where α is the path loss exponent which depends on environment ($2 \leq \alpha \leq 5$). Let β be the receiver sensitivity threshold such that signal is properly decoded at the receiver if $P_r \geq \beta$. The communication range of a node is the distance at which $P_r = \beta$ in absence of any other interference. Then the power level of nodes can be presented in terms of their communication range. As an example, in G_C all nodes are operating at power level $P(d_{min})$ which is necessary and sufficient to achieve communication range of d_{min} at all nodes.

From this minimum, the power level capability of each node is uniformly increased. As a result, every node increases its communication range by a factor of f from the COMPOW range. This is achieved by tuning its power level to $P(f \cdot d_{min})$. This way, increase of power levels are normalized to the COMPOW range (d_{min}), not to the COMPOW power level ($P(d_{min})$). Pathak and Dutta (2011) refer to f as the *growth factor* of connectivity.

Pathak and Dutta (2011) study the effect of increasing power level capability on the throughput capacity of the network. The throughput capacity of the network is studied for uniform node-to-node traffic pattern, for both uniform and clustered topologies, using simulation. Since the TDMA-based link scheduling problem is known to be NP-hard for protocol interference model (Draves et al. 2004), K-hop

interference model (where $K \geq 2$) (Sharma et al. 2006) and SINR-based physical
interference model (Goussevskaia et al. 2007), they adopt in their simulation a greedy
link scheduling approach similar to that previously used in literature (Brar et al. 2006;
Pathak et al. 2008) that generally performs within a constant factor of the optimal
schedule. Briefly, the end-to-end traffic demand between nodes is represented using
a traffic demand matrix, which can be converted to a per-link transmission demand
matrix once some routing has been determined for each source-to-destination node
pair. Links are sorted by an interference score; such as the number of other links it
interferes with. The greedy scheduler chooses links greedily from this list that can
be scheduled together, until no more links can be added to the current scheduling
slot. Then a new slot is created and the process repeated, until no more transmission
demand remains. In what follows, the data is routed on a shortest path which is
closest to the straight line connecting the source and the destination [e.g. straight
line routing (Kwon and Shroff 2007)].

For uniform topology, Pathak and Dutta (2011) propose an approximate theoretical model of the capacity as well. This model is based on observing that. as power
capability goes up, the average transmission link length in the network goes up, and
the number of double-disks of interference that can be packed into the network area
goes down; this models the effect of lowered spatial reuse. At the same time, the
total number of hops required to satisfy the traffic load goes down. Figure 3.12a, b
shows results of these attempts to approximate the optimal throughput theoretically.

Figure 3.13a shows the result of combining these two observations to achieve a
bound on the throughput capacity of the network, compared with simulation results;
they agree in general nature but are noticeably different, and disagree in the trend at
higher growth factors (the model keeps rising but the simulation flattens out). Apart
from the overall sub-optimality introduced in the simulation results due to the greedy
scheduler, the disagreement in the trend may possibly be due to the dominance of
loner links. If a link can not be scheduled with any other link in the network, it is

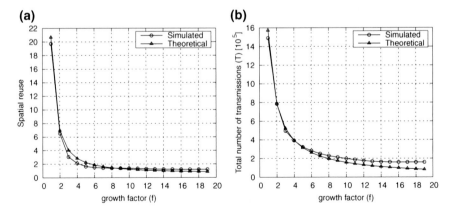

Fig. 3.12 Approximate theoretical model of (**a**) spatial reuse, (**b**) number of transmissions required
in schedule with increasing growth factor

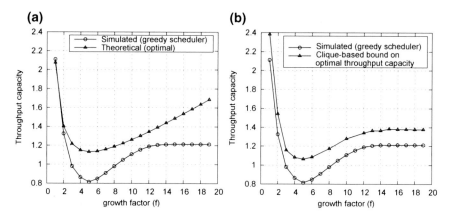

Fig. 3.13 Comparison of simulation results with (**a**) combined theoretical model from Fig. 3.12a, b, (**b**) simple conflict clique based bound, for network throughput capacity

referred as a loner in its slot. Long links may get created near the center, which forces the spatial reuse to drop to one. Pathak et al. (2008) characterized the length of such links, and showed that their existence is independent of the scheduling algorithm since even a theoretically optimal scheduler can not achieve spatial reuse of more than one while scheduling a loner link. When power increment of nodes reaches a certain high level ($f \geq 14$), many (upto 70 %) of the links become loners. Such considerations can be used to obtain a different estimate (actually, a bound) on the throughput. If the vertices of the conflict graph are weighed with the traffic traversing the corresponding network links, then a maximum weight clique in the conflict graph identifies a set of links that must be each scheduled in a different slot, providing a lower bound on optimal schedule length (upper bound of optimal throughput). Figure 3.13b shows this bound in comparison with, again, the simulation results.

Pathak and Dutta (2011) also study random topologies (generated by homogeneous Poisson point processes) and clustered topologies [using Matérn cluster process (Illian et al. 2008)]; their results show that random placement can be considered a point on the continuum between grid and clustered topologies from the point of view of throughput capacity. Most real-world topologies actually display some degree of clusteredness. They are either highly clustered, or fall between random topologies and clustered topologies; commonly referred to as diffuse clusters. Since the COMPOW range is the minimum common range such that the nodes are connected, it increases as the topology becomes more and more clustered. This results in higher intra-cluster connectivity but inter-cluster connectivity still remains low. In such case, links providing connectivity between the clusters become traffic bottlenecks under N2N traffic. These links require large number of transmissions and consume large number of time slots when link scheduling is performed. This increases the overall schedule length and reduces the attainable capacity.

When power is increased uniformly, new inter-cluster links are added which share the traffic burden and reduces the traffic bottlenecks in the network. This results into higher spatial reuse among the links and overall throughput capacity increases. We previously referred to CLUSTERPOW and similar power control techniques that can be also seen as utilizing this effect. Increasing power levels of nodes in such cases actually has uniformly positive effect on throughput. This is surprisingly different from uniform topologies where increasing power levels in N2N traffic had mix effects on throughput. Pathak and Dutta (2011) studies the effect of clusteredness by focusing on the graph clustering coefficient proposed in Newman et al. (2002), also known as *transitivity* of a graph, which is proportional to the ratio of the number of *triangles* (subgraph induced by three vertices that has three edges) to the number of *triples* (subgraph induced by three vertices that has two edges). By simulating clusters of different intensities, Pathak and Dutta (2011) show that the transitivity of the COMPOW graph of the topology (which is typically higher for more clustered graphs) shows an interesting relation with how throughput changes when power capability is increased. The characteristic observed for the grid case is pulled down as an effect of increasing clusteredness as expected, but the lower growth factor region is affected significantly more than the higher growth factor region; so much so that for topologies which have transitivity similar to diffuse cluster or higher, the curve becomes monotonically increasing. Thus for such topologies increasing power levels always results into better throughput. This is observed both in simulated results as well as the clique-based bounds. This is a practically useful result because, for a given set of nodes, transitivity of the COMPOW graph can be calculated and then it is possible to predict the behavior of throughput with increase of power levels.

3.3.2.4 The Case of Node-to-Gateway Traffic

A commonly used and practically important traffic pattern is one in which traffic originates at all nodes, but are all destined for a particular gateway node. The most representative use of such traffic pattern is observed in Wireless Mesh Networks (WMNs) in which all nodes send traffic to one or more gateway nodes which forwards it further to the Internet. In one of the most useful results about capacity of WMNs, Jun and Sichitiu (2003) proved that per node throughput in WMNs can not be more than $O(1/n)$ which is significantly worse than classical result of $O(1/\sqrt{n})$ presented in Gupta and Kumar (2000). This is because traffic of every node, no matter how many hops away it is from the gateway, has to ultimately traverse through the bottleneck links connecting to the gateway. Per node throughput of $O(1/n)$ is also achievable in WLANs, but it has been empirically observed that WMNs often achieve even worse throughput than WLANs (often by a factor 2 or 3) (Fig. 3.14).

It was shown by Pathak and Dutta (2011) that WMNs in fact achieve per node throughput of $O(1/\delta n)$, where $\delta \geq 1$ is a factor dependent on the hop-radius of the network graph. *Hop-radius* is defined as length of the longest path from a node to the gateway assuming that the gateway has the least eccentricity among all nodes. The result pertains to networks where interference model is based on the distance. The theoretical derivation of this result depends on recognizing that the nodes of the network can be divided into *tiers* depending on their hop distance from the gateway,

Fig. 3.14 For Node-to-
Gateway traffic, almost all
links of the first three tiers
form a bottleneck collision
domain

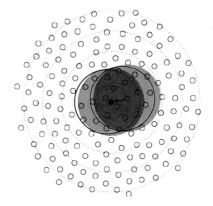

Fig. 3.15 Network throughput for N2G traffic: sin-
gle gateway, model and
simulation

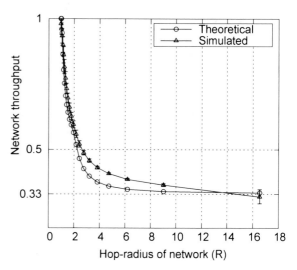

and the many links (pessimistically, all links) of the three tiers closest to the gateway
fall in the same bottleneck collision domain. The insight provided by Pathak and
Dutta (2011) is that these tiers also restrict the minimum length of the schedule,
and thus determine the throughput of the network. Thus the Surprisingly, factor δ
converges to 3 as $R \to \infty$ in (4). This suggests that no matter how large the hop-
radius of the network becomes, WMNs can still achieve one third of the capacity of
WLANs even in the worst case. This is due to the fact that at least one link in first
three tiers can always be scheduled in almost all slots. (It is important to note that
the value to which the δ converges also depends on how *BCD* is calculated, which
in turn is dependent on the interference model under consideration.)

This prediction is verified by simulation as shown in Fig. 3.15. These results
suggest that it is always better to increase power level of nodes which decreases the
network hop-radius and increases the throughput. This is a useful capacity result

especially in case of WMNs since it proves that reducing worst case hop distance to gateway always performs better in terms of throughput.

In large metro-scale WMNs, having one gateway is not scalable in meeting the demand. When more and more nodes are assigned the role of gateway, network throughput increases but associated cost also increases since such gateway nodes typically have a high-speed fiber link or a satellite connection. In such case, the objective of network designer is to achieve as much throughput as possible with fewer number of gateways. A similar result to the above is shown to hold true when the WMN utilizes multiple gateways, instead of a unique one, by Pathak and Dutta (2009). The problem of placing k gateways in the network can be formulated as a facility location problem, or a problem of finding k-medians in the network graph, which are known to be NP-hard. Pathak and Dutta (2009) use a common but effective heuristic gateway placement by identifying clusters in the topology, or simply by dividing the network into a square grid when the topology is close to uniform. The general nature of their results is as shown in Fig. 3.16. When nodes double their transmission range, on an average 20 % more throughput can be achieved for any given number of gateways (that is, the characteristic line moves 20 % up). This is a cost-effective solution since better throughput can be achieved with lesser number of gateways when nodes increase their power levels. After a certain value of growth factor f, further power control advantages become insignificant, but these limiting values of f can be quite high when the COMPOW range is small.

3.3.2.5 Energy Efficiency

Some power control mechanisms aim to minimize the overall power consumption in traditional ad-hoc networks. With increasing outdoor deployments of wireless mesh

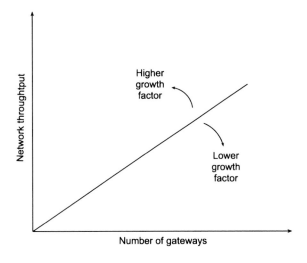

Fig. 3.16 Network throughput for N2G traffic: nature of results for multiple gateways

networks, there is an opportunity to utilize alternate sources of energy, like solar energy, to operate the wireless routers. Energy consumption, though not primary in WMNs, remains an important objective of any power assignment (or topology control) mechanism. Jia et al. (2005) propose several approximation algorithms for finding a power assignment of nodes in wireless ad-hoc network such that the topology graph is k-connected (i.e., it remains connected upon removal of fewer than k vertices) and the total power utilized is the lowest. Wireless links often display unpredictable behavior and hence such fault tolerant topology holds importance in terms of survivability in WMNs. As defined in Jia et al. (2005), an i-nearest-neighbor subgraph of G is a spanning subgraph of G in which there is an edge between two nodes u and v if and only if either u is one of the i nearest neighbor of v in G, or vice versa. A subgraph F of G is called a k-connectivity augmentation to H if $H \cup F$ is a k-connected spanning subgraph of G. The algorithm of Jia et al. (2005) first constructs the $(k - 1)$-nearest neighbor graph $G(k - 1)$ from the maximum connected topology and then finds a k-connectivity augmentation F to $G(k - 1)$ and outputs $G(k - 1) \cup F$ with the desired property.

Multiple coverage solar powered 802.11 mesh networks with load balancing are considered by Vargas et al. (2007); two algorithms are provided that try to dynamically activate and deactivate mesh APs based on current traffic demands.

3.3.3 Dynamic Power Control

With dynamic power control strategies, every node changes its power level for transmission frequently over time. Such changes can be made on a per link, per destination, per TDMA slot or per packet basis.

Many proposed mechanisms perform power allocation locally at every node based on the current condition of its neighbors. The PATE (Power Assignment for Throughput Enhancement) algorithm presented by Xiong et al. (2003) is one such approach; it tries to avoid congested neighbors by choosing next-hop nodes which are less loaded. Power assignment is performed in such a way that connectivity of the network is maintained while the least congested neighbor that will create low interference to other nodes is chosen. A cost function is presented to determine the neighbors, and the corresponding required power levels to reach them. The study of Monks et al. (2001) proposes the Power Control Multiple Access (PCMA) MAC protocol where the transmitter chooses the transmit power level based on how much interference the receiver can tolerate. It uses a separate control channel to send "busy tone signals" which advertise the tolerance levels. Similarly, Power Controlled Dual Channel (PCDC) (Muqattash and Krunz 2003) uses a separate control channel for advertising the interference tolerance. Though both PCMA and PCDC result in increased throughput due to informed decisions regarding power control, both assume wireless devices can transmit and receive at the same time on control and data channels; this requires an additional radio for each communicating device. This limitation was addressed by Muqattash and Krunz (2005), who also proposed an improved power control protocol; POWMAC uses an access window to allow for a series of RTS/CTS exchanges to take place before several data transmissions can take place concurrently. The received signal strength is then used to dynamically bound the transmission power of potentially interfering terminals in the vicinity of a receiving terminal. The required transmission power of a data packet is computed at the intended receiver, to allow for some interference margin at the receiver. This will allow multiple transmissions to take place concurrently in the neighborhood. Though POWMAC achieves concurrent transmissions using one channel only, sometimes contention may occur during the access window.

The problem of power assignment becomes even more complex when decisions must be made in a distributed fashion, with limited information available locally. To simplify the design, Akella et al. (2007) proposes a feedback based fall-back power control algorithm. If a pre-determined number of transmissions are successful, the sender decrements the transmission power level until it reaches the lowest required power level without affecting the intended data rate. While decreasing power, if it encounters data loss at a certain level; it starts incrementing power level until it recovers the data rate (or reaches the highest possible power level). With the objective of power allocation in a decentralized network, Sharma and Teneketzis (2007) formulate power allocation as a problem in which interference (caused by transmissions of other users) is viewed as an external emergent condition by nodes, quantified as an "interference temperature". The approach presented tries to satisfy the interference temperature constraints and maximize a function of the sum of users' utilities.

In the IEEE 802.11 MAC standard, whenever a node wishes to transmit, it first senses the medium; if the sampled signal strength is below the carrier sense threshold, it initiates the transmission. The value of the threshold and transmit power together dictate when and at what power level a node transmits. All combinations of these two quantities may not be useful; Fuemmeler et al. (2006) argue that the product of transmit power and carrier sense threshold should be a constant for each transmitter in the network. Based on such a mechanism, nodes transmitting at large power levels should use lower carrier sense threshold because they cause more interference to others and should be more careful before initiating their transmission. As in Kim et al. (2006b), the spatial reuse can be increased in wireless multi-hop networks such as mesh by either reducing transmission power, or increasing the carrier sense threshold. The study of Kim et al. (2006b) shows that there is a trade-off between the level of spatial reuse and the data rate that can be sustained by a transmission. It also shows that if the achievable channel rate is a continuous function of SINR, network capacity depends only on the ratio of transmit power to the carrier sense threshold. When the set of channel data rates available are discrete (as in realistic protocols like 802.11), tuning the transmission power while keeping the carrier sense threshold constant can have more advantages. Finally, they provide a localized power and rate control algorithm, where the transmitter monitors the current interference, and determines the power level accordingly. The algorithm chooses a power level in such a way that the sender can sustain the maximum possible data rate, and the interference caused by it to other nodes is minimum. Similarly, Yang et al. (2007)

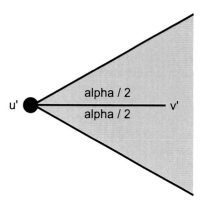

Fig. 3.17 The use of directional antennas to restrict transmission to a cone

conclude that higher throughput can be achieved if the area silenced by the transmitter is reduced as much as possible, as long as it covers the interference area (containing the set of nodes that would cause collision at the receiver, if they also transmitted) of the receiver (Fig. 3.17).

3.4 Topology Control

Topology control mechanisms try to choose a certain set of links to be used out of all possible links in a network for a certain specific objective like lower power consumption, higher throughput, better fault tolerance etc. In WMNs, topology control can be used for reducing interference and thereby reducing MAC collisions. The studies of Li et al. (2001, 2005) and Wattenhofer et al. (2001) present the Cone-Based Topology Control (CBTC) algorithm where each node finds a minimum power level at which it can reach some neighbor in every cone of degree α. Such a topology is shown to be preserving connectivity when $\alpha < 5\pi/6$. The objective is to reduce overall energy consumption with increased throughput. Though energy conservation is not a major objective for power or topology control in mesh, such an approach still holds importance as it improves on throughput and preserves network connectivity. Such topology control mechanisms are especially useful when mesh nodes employ directional antennas to reduce interference. As shown by Li et al. (2001, 2005) and Wattenhofer et al. (2001), cone based topologies can significantly reduce the edge density, and generates a sparser topology while maintaining a robust overall connectivity (see Fig. 3.18).

Ramanathan and Rosales-Hain (2000) formulate the topology control problem as a constrained optimization problem for (bi)connectivity while optimizing the maximum power used per node. Two centralized spanning tree based algorithms (CONNECT, BICONN-AUGMENT) and two distributed heuristics (LINT, LILT) are presented by Ramanathan and Rosales-Hain (2000) that attempt to achieve connected topology with minimum power utilization.

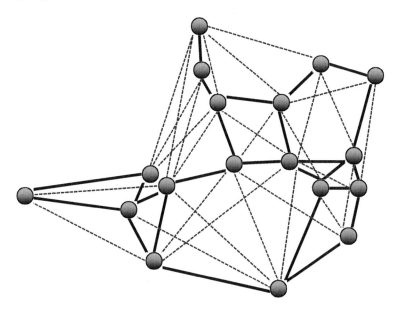

Fig. 3.18 The cone-based topology (*bold lines*) is sparser while maintaining robust connectivity than the omni-directional topology (*fine lines*)

Most power/topology control approaches focus on achieving sparser topologies for higher throughput, without explicitly considering the underlying issue of interference. The study of Burkhart et al. (2004) disproved the common belief that sparseness (lower node degree) of topology invariably achieves lower interference. It defines network interference in terms of maximum value of *coverage* of any link *uv* (the number of nodes affected by communication on this link when *u* transmits at the power level required to reach *v*). It shows by examples that topology control based on nearest neighbor or graph planarity (a graph is planar if it can be drawn on a plane without any edges crossing) cannot guarantee interference-optimal topologies. Using an earlier definition of network interference, Burkhart et al. (2004) proposes an MST-based centralized algorithm (Low Interference Forest Establisher or LIFE) to determine the minimum-interference connected topology. The Minimum Spanning Tree (MST) is generated by selecting links with lower coverage values which ultimately reduces the overall network interference. It also proposes a distributed variant of LIFE to locally obtain a minimum interference spanner topology. Similarly, Moaveni-Nejad and Li (2005) present algorithms to find topologies with lower *average* link/node interference.

Interference defined on the basis of link coverage as by Burkhart et al. (2004) is constrained to be sender-centric, in that it does not account for receivers that may also be interfered with when a link *uv* is active. The study of von Rickenbach et al. (2005) extends the definition by specifying that interference of a node *v* represents the number of nodes covering *v* with their transmission range disk when reaching their farthest neighbor. Based on this interference model, it proposes an approximation algorithm to yield minimum interference connected topology in the so-called *highway model*. Topology formation by Burkhart et al. (2004) and Moaveni-Nejad and Li (2005) accounts for per link interference only while building the low interference MST graph. This might lead to very high interference when end-to-end multi-hop routing paths between the node pairs are considered. Blough et al. (2005) provide a topology control algorithm where a link *uv* is chosen in the topology if and only if it belongs to a minimum interference path connecting any nodes *w* and *z*.

It is natural to look to topology control to address the unique characteristics of directional antennas, when these are used. Kumar et al. (2006b) introduce a topology control mechanism for mesh nodes with directional antennas. The k-degree spanning tree algorithm finds directions for k directional antennas at every node in such a way that the node has at least k incident edges. The topology control algorithm presented by Li et al. (2003c) and Khan et al. (2009) try to determine power levels for nodes such that under the case of random node failures, the remaining topology retains k-connectivity with a high probability, over a longer period of time.

3.4.1 Graph Planarity

It has long been generally acknowledged that network topologies that are planar graphs on an Euclidean embedding of the underlying node placement are highly likely to demonstrate high energy efficiency and spatial reuse. We next mention a few planar graphs which have been especially proven useful in the context of wireless networks.

A Euclidean complete graph contains a set of vertices and their Euclidean coordinates. It also contains a set of edges where there exists an (undirected) edge between every pair of vertices, and the weight of an edge is the Euclidean distance between its end points.

3.4.1.1 Minimum Spanning Tree

A Euclidean minimum spanning tree (MST) is a graph containing all vertices with edges chosen such that the total weight of all chosen edges is the minimum. The most fundamental advantage of MST is that its node degree is upper bounded by 6. On the other hand, the disadvantage of using MST in ad-hoc networks is that it cannot be generated using locally available information; this requires a large number of control messages to be exchanged between nodes.

A localized version of MST was proposed by Li et al. (2003a), as previously described in Sect. 3.3.2.2.

3.4.1.2 Relative Neighborhood Graph

In a Relative Neighborhood Graph, nodes x and y are connected if and only if the area of intersection of two circles with radius equal to the distance d_{xy} centered at x and y does not contain any other node. It is known that such a graph does not have a bounded node degree.

3.4.1.3 Gabriel Graph

A Gabriel Graph is generated by connecting two nodes x and y if the disk where edge xy is a diameter does not contain any other node. It is known that similar to the Relative Neighborhood Graph, the Gabriel Graph also does not have a bounded node degree.

3.4.1.4 Delaunay Triangulation Graph

This graph is built by connecting two nodes x and y if the circum-circle of the triangle developed on x, y and any third node z is does not contain any other node. It is not possible to build such a triangulation locally without the availability of global information. Other variants of Delaunay Graph have been proposed by Gao et al. (2001) and Li et al. (2003b, 2004).

3.4.1.5 Hypocomb Graph

In recent research, Li et al. (2011) develop a family of connected planar graphs which have a special importance in wireless networks. Three graphs belonging to the Hypocomb family are Hypocomb, Reduced Hypocomb and Local Hypocomb. Hypocomb and Reduced Hypocomb graphs are subsets of the complete graph, while the Local Hypocomb is a subset of the Unit Disk Graph.

For a given set of nodes and their Euclidean coordinates, a Blocked-Mesh graph can be constructed by drawing rays in four directions from each vertex, and blocking them when they meet each other. A Hypocomb graph is extracted by linking vertices that have their rays blocked by each other in Blocked-Mesh graph. The Hypocomb graph has an unbounded degree. In reduced Hypocomb, two vertices have an edge between them if and only if their rays block mutually. The advantage of Reduced Hypocomb is that it has a bounded degree of 6. The last and the most important graph of the family is known as the Local Hypocomb graph. In this graph, any edge that belongs to the Unit Disk Graph but does not belong to the Reduced Hypocomb graph is removed. The advantage of the Local Hypocomb graph is that it is generated using only local neighborhood information and has its degree bounded to 8.

Li et al. (2011) go on to investigate these methods by studying the graphs built by these methods on a given specific node placement. Their results show that considerable variation in the graph can arise even for a small set of nodes, due to the specifics of the approaches. They also studied the mean and maximum node degrees of the nodes in the graphs constructed as a function of the number of nodes. They show that for all approaches, the mean degree increases nearly linearly with the number of nodes, with similar mean degree for each approach, though some like RHC tend to flatten out after about 200 nodes. Most of the approaches show a different trend with respect to the maximum degree, which tends to quickly increase to a maximum value before 50 nodes, then stays nearly level for graphs with larger numbers of nodes. However, RHC and LHC show the characteristic of reaching a smaller maximum degree than the Hybocomb (as well as previous approaches such as Delaunay and Gabriel graphs), which are also reached somewhat earlier; this could be a desirable characteristic under some design goals.

3.5 Scheduling

Link scheduling estimates the interference conflicts between the links having transmission demands (based on the interference model) and tries to achieve a conflict-free feasible transmission schedule. The first generation of scheduling algorithms (Gandham et al. 2008; Hajek and Sasaki 1988; Moscibroda and Wattenhofer 2005; Ramanathan and Lloyd 1993; Ramaswami and Parhi 1989) for multi-hop wireless networks were based on simplified graph models. Such algorithms mainly followed characteristics like the network topology graph and often failed to capture the issues of dynamic wireless medium such as interference. The study of Grönkvist and Hansson (2001) indicated that the graph-based scheduling does not take full interference knowledge in account while performing the link scheduling. It might be too optimistic by allowing few unintended transmissions nearby the receiver which may cause collisions or can be too pessimistic by not allowing such a transmission which can cause tolerable interference at the receiver. Compared with the physical-model type SINR-based scheduling, it achieves lower network performance. Along the same line, Behzad and Rubin (2003) conclude that transmission scheduling based on maximal independent set in graph-based interference model may suffer from intolerable SINR at the receivers, yielding low network capacity. Even maximizing the cardinality of the independent sets does not yield any better performance. Similarly, Moscibroda et al. (2006) proved with theoretical examples and experimentation that such graph-based models can undermine the achievable capacity even for simple settings of the network. They conclude the need of protocol design based on more realistic SINR-based physical interference model.

CSMA-CA and TDMA are two MAC protocols commonly used in the wireless networks. Both the protocols have their pros and cons which makes them viable choice for WMN MAC. CSMA is a simple, robust and scalable medium access technique. It does not require any time synchronization and, addition or removal of nodes from the network can be handled in distributed fashion. The Distributed Coordination Function (DCF) of 802.11 is an implementation of CSMA with binary exponential back-off. In DCF, a node wishing to transmit first senses if the medium is busy or not. If the medium is not busy, the node proceeds with the transmission but if the medium is found busy, node chooses a random back-off time and waits for that duration until the next retry. Since such carrier sense only works among one-hop neighbors, transmissions of two nodes which can not listen to each other can collide at the receiver. Such a problem is typically referred as *hidden terminal problem* and is a critical problem with 802.11 MAC. RTS/CTS (Ready to send/Clear to send) are two messages which are used to alleviate the problem but they themselves incur higher overhead. On the other hand, TDMA does not suffer with MAC collisions in its ideal implementation because each node only transmits in its dedicated slot which does not conflict with its interfering nodes. When traffic is relatively stable (non-sporadic), TDMA can achieve maximum system capacity but there are several issued with TDMA too. Its distributed implementation is substantially difficult and requires tight time synchronization. Also, it is relatively inflexible to dynamic changes to the topology.

Since interference can be caused by many nearby nodes in mesh networks, medium access and link layer protocols are much more complicated to design. CSMA-CA based MAC protocols often suffer from lower throughput in multi-hop mesh due to its conservative design but still offers advantages of its distributed nature and standardized implementation (802.11). On the other hand, TDMA based MAC protocols are known to be more efficient due to their work-conserving nature which is better suitable for relatively stable traffic pattern of mesh backbone. The problem with TDMA based MAC protocols is that their actual implementation requires thorough engineering efforts, which is often outside the scope of research. Due to this reason, TDMA scheduling protocols can be classified into coarse-grained and fine-grained protocols. In coarse-grained TDMA protocols, emphasize is given to link scheduling with various valid assumptions of interference model, traffic demands and centralized control. To realize their potential in practice, they often have to depend on existing link layer technologies for framing, link layer acknowledgments etc. while handling medium access and transmission control at upper layers. While fine-grained TDMA protocols often handle all link layer functions at MAC layer, which makes them increasingly difficult to implement in practice. For brevity, we do not distinguish between the two kinds of TDMA protocols and discuss them together next.

3.5.1 TDMA-Based Link Scheduling Protocols

Recently, CSMA-CA is shown to be not suitable for multi-hop wireless networks because of its conservative medium access and hidden/exposed terminal problems. On the other hand, TDMA based link scheduling can achieve better spatial reuse in case of WMNs where traffic demand between routers are assumed to be relatively stable. Along the first step towards designing realistic scheduling protocols, Grönkvist et al. (2004) provided LP formulation for node-based and edge-based spatial reuse TDMA scheduling for physical interference model. The study of Grönkvist (1998) provided traffic controlled schedule generation algorithm but computational complexity of the approaches of Grönkvist (1998) and Grönkvist et al. (2004) can be of a high order.

It is important to model interference relationship between links based on respective interference model before they can be scheduled. Problem of link scheduling can be represented as problem of finding maximum independent set in the conflict graph. Vertices connected to each other in the conflict graph represent those links of communication graph which interfere with each other and cannot be scheduled simultaneously. The study of Jain et al. (2003) first designs conflict graph for protocol interference model indicating which set of links interfere with each other and cannot be scheduled together. Conflict graph in physical interference model has vertices which correspond to edges in communication graph. There is a directed edge between two vertices (edges in communication graph) whose weight indicates what fraction of the maximum permissible noise at the receiver of one link by activity on another link. This conflict graph based on interference model adds interference constraints to the LP formulation which optimizes the throughput for single source-destination pair. The LP formulation requires calculating all possible transmission schedules and it is shown to be computationally expensive.

To avoid the complex edge-based conflict graph of Jain et al. (2003), a method is proposed by Brar et al. (2006) to simplify the design of conflict graph in the physical interference model. The node-based conflict graph is designed by keeping the vertex set same as the communication graph and adding a directed edge uv between vertices u and v whose weight corresponds to the received power at v from of the signal transmitted by u. The only constraint in this case is that a node cannot transmit and receive on different links simultaneously. So, feasible schedule of links in physical interference model forms a *matching* in communication graph and should comply with SINR constraint. With non-uniform link demands and uniform random node distribution, Brar et al. (2006) provide computationally efficient polynomial-time scheduling algorithm for which an approximation factor relative to the optimal schedule has been proved. The algorithm is not distributed and still requires a central entity to perform schedule calculation. The computational complexity of spatial TDMA scheduling is known to be very high especially when using the physical interference model. In such cases, it becomes increasingly difficult to estimate or even bound the optimal scheduler performance and compares it with the proposed strategy. The study of Björklund et al. (2004) derives a column generation method using *set covering* formulation which efficiently solves the scheduling problem. The method is also used to derive tight bounds on the optimal scheduling performance which can be very useful as a benchmark for performance comparison.

Similar algorithm for double disk based interference model is presented by Pathak et al. (2008). The end-to-end traffic demand between nodes is represented using a traffic demand matrix (T_R). Once the shortest path routing is performed, T_R yields per-link transmission matrix (T_X). We assume that there is a central controller entity which performs link scheduling. In the operation of greedy scheduler, first all links of T_X are sorted based on their interference score. Interference score of a link is the number of other links with whom the given link interferes and hence can not be scheduled simultaneously. Then scheduler chooses the first link in order to be scheduled in the current slot and tries to add more and more non-interfering links greedily until no more links can be added to the slot. The procedure repeats until all transmission requests of T_X are satisfied. Brar et al. (2006) showed that such a scheduler has the time complexity of $O(m \cdot n \cdot T)$, where $T = \Sigma_{i=0}^{n} \Sigma_{j=0}^{n} T_{X_{ij}}$. If the total offered load $G = \Sigma_{i=0}^{n} \Sigma_{j=0}^{n} T_{R_{ij}}$ and greedy scheduler requires S slots to schedule all the links, the network throughput is G/S traffic units per unit time.

3.5.1.1 Impact of Radio Propagation Model and Interference Model

It is clear that schedule length and the resultant throughput capacity is dependent on how the RF signal propagates and how the interference relationship between links are modeled. Using a TDMA greedy scheduler (similar to the one described above), Stuedi and Alonso (2007) evaluated the relationship between the schedule length, throughput capacity, interference model and RF propagation model.

As we discussed above, physical interference model is often preferred in studies due to its ability to realistically model the interference, but it introduces significant complexity in other protocols at other layers (such as scheduling). This has led to increasing number of research studies adopting protocol interference model. Thus, it is necessary to understand the actual difference in throughput capacity they yield when used with a scheduling strategy. It was shown by Stuedi and Alonso (2007) that utilizing these two models when scheduling using basic network settings in fact yields a significant qualitative and quantitative difference in throughput capacity.

There is an additional dimension in modeling that is required to be considered—radio propagation. As we saw before, there are two different kinds of radio propagation models largely used in research—path loss radio propagation with and without log-normal shadowing. Formally, as shown by Rappaport (2001), for a transmitting node u, the received signal strength at receiver v at a distance d_{uv} is modeled as:

$$PR_{dBm}^{v}(d_{uv}) = PT_{dBm}^{u} - PL_{dB}(d_0) - 10\eta \log_{10}\left(\frac{d_{uv}}{d_0}\right) + S_{dB}^{\sigma} \qquad (3.2)$$

where PT_{dBm}^{u} is the transmit power, $PL_{dB}(d_0)$ is the reference path loss at a distance d_0, η is the path loss exponent and S^{σ} is a zero-mean Gaussian variable with standard deviation σ. A wireless link exists between u and v if the received power $PR_{dBm}^{v}(d_{uv})$ is not less than some given minimum threshold P_{min}. Such a threshold is typically referred as the *receiver sensitivity*. Such a model of RD propagation is known as path loss radio propagation with log-normal shadowing. On the other hand, when the shadowing factor ($\sigma = 0$), the resultant model becomes a path-loss model without shadowing. We refer to path loss model without shadowing as just path-loss model, and path-loss model with shadowing as shadowing model.

Figure 3.19 shows the throughput capacity using greedy scheduler with different interference models under path-loss propagation. The capacity decreases with the

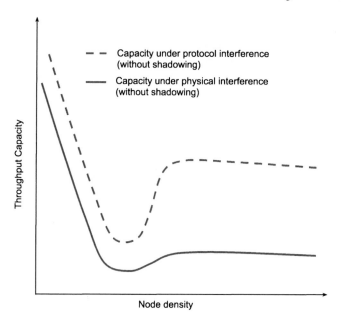

Fig. 3.19 Network throughput (under greedy scheduler) behavior depends on the interference model used

increase in network density due to increase of interference [also known from the work of Gupta and Kumar (2000)]. Other behavior that is worth noting is that the capacity curve follows a three-phase transition in which at first the throughput decreases, then increases which is then followed by slower decrease again. This is because a large number of nodes become connected with increase in density which allows more and more flows to route their data to their destinations. This initially increases the capacity which is later decreased because the negative effect of increased density becomes prominent in this later part of curve with high node density. Other fact that is noticeable is that physical interference model leads to lower spatial reuse as compared to protocol interference model which in turn results into lower throughput capacity.

Now when shadowing effect is introduced for RF propagation, the resultant throughput capacity is shown in Figs. 3.20, 3.21. As it is shown in both the figures, higher value of shadowing results into increased throughput capacity. This is due to the fact that increase in shadowing phenomena reduces the number of transmission that required to be scheduled in order to satisfy the traffic demand. The increase is observed in both protocol (Fig. 3.20) and physical (Fig. 3.21) interference models.

Finally, Fig. 3.22 summarizes the capacity results of protocol interference model with path-loss propagation and physical interference model with shadowing interference model. This yields useful insights on how conventional interference and RF propagation models can give surprisingly different performance results as opposed to more realistic models.

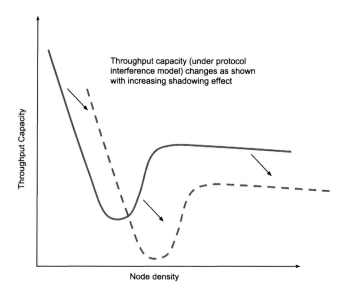

Fig. 3.20 Shadowing shifts throughput characteristic under protocol interference model (compare with Fig. 3.19)—overall throughput is lower but onset of throughput reduction with increasing node density is delayed

Fig. 3.21 Shadowing modifies throughput characteristic under physical interference model (compare with Fig. 3.19)—throughput reduction is less drastic at higher node densities

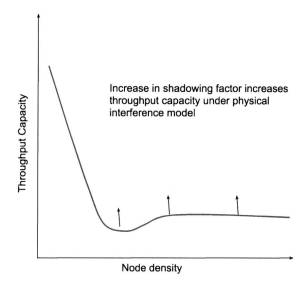

3.5.1.2 Enhancements Using TDMA-Based MAC Protocols

To enhance the performance of TDMA, Djukic and Valaee (2007b) consider a problem of designing minimum delay schedules via intelligent ordering of link transmissions in TDMA MAC which ensures lower node-to-gateway delays. For example, if

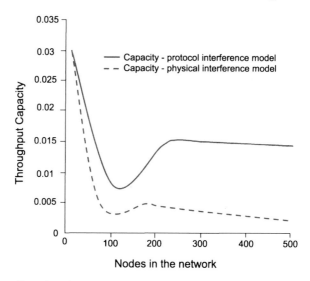

Fig. 3.22 The effect of realistic models: physical interference model with shadowing produces divergent characteristic from that produced by protocol interference and path-loss model

outgoing link is assigned the slot before the incoming link in a TDMA frame then end-to-end delay may become significantly high. Instead Djukic and Valaee (2007b) formulate the TDMA scheduling problem as a network flow problem on the conflict graph, solution to which minimizes the delay on a routing tree rooted at the gateway. First, low delay transmission ordering of links is found and then using it with the link conflict information, Bellman-Ford algorithm is used to find feasible TDMA schedule in polynomial time. The study of Djukic and Valaee (2007a) extends the work of Djukic and Valaee (2007b) by providing a distributed scheduling algorithm. It is first shown that TDMA scheduling problem is equivalent to finding shortest paths in augmented partial conflict graph which is available at nodes based on their local information. Using distributed Bellman-Ford algorithm, conflict-free and feasible schedule can then be derived.

Cross-layer optimization problem has also attracted researchers to derive resource allocation solutions for multi-hop wireless networks. The seminal work of Tassiulas and Ephremides (1992) first addressed throughput-optimal scheduling. It showed that scheduling mechanism is throughput optimal if it maximizes queue-weighted sum of rates and also characterized maximum attainable throughput region. Scheduling policy proposed by Tassiulas and Ephremides (1992) is centralized and suffers from a higher computational complexity. The study of Lin et al. (2006) showed that relaxing scheduling component in cross-layer design can actually open up many chances for new distributed, simpler and provably efficient algorithms. The imperfect scheduling (also known as greedy scheduler or maximal weight scheduler) policy determines the schedule by choosing links in decreasing order of the traffic backlog at every node. As described by Joo et al. (2008), such greedy maximal scheduler often performs near optimal empirically but the known bounds of its performance are still very loose. It is known that such maximal scheduler is guaranteed to achieve at least half of the maximum throughput region for node exclusive interference model (Lin and Shroff 2006). Such efficiency ratio is shown to be dependent on *interference degree* of the network by Chaporkar et al. (2005). It is shown that with bidirectional equal power geometric (double disk) interference model, such scheduling can achieve $1/8$ of maximum throughput region. Similarly, it was shown by Sharma et al. (2006) that when $K \geq 2$, greedy maximal scheduler can achieve the efficiency ratio of $1/49$. The study of Dimakis and Walrand (2006) showed that network topologies that satisfies local pooling condition can achieve maximum throughput in case of longest-queue first scheduling. Using this results, it was proven by Joo et al. (2008) that greedy maximal scheduler can achieve full system capacity in tree networks under K-hop interference model and has efficiency ratio between $1/6$ to $1/3$ in geometric network graphs. Such an imperfect scheduling (Lin and Shroff 2006) has led way to many joint algorithms for scheduling (Gupta et al. 2007), congestion control (Sharma et al. 2007), channel assignment (Lin and Rasool 2007) and routing (Lin et al. 2007). A good survey for such approaches is by Lin et al. (2006).

One interesting extension of the TDMA scheduling problem is to design collision-free link scheduling of the broadcasts. Network wide broadcasting of messages is one fundamental operation in ad-hoc networks and several upper layer protocols depend on such functionality. As outlined by Gandhi et al. (2003), broadcast scheduling with link interference conflicts incurs a latency which is calculated as duration between time of first broadcast and time at which all nodes receive the broadcast. The objective is to compute a broadcast schedule which requires lesser number of slots (minimum latency) and fewer numbers of retransmissions. The study of Gandhi et al. (2003) first proved that minimum latency broadcast scheduling is also NP-hard and provided approximation algorithm for it. The latency of approximation algorithms was subsequently improved by Cicalese et al. (2006), Elkin and Kortsarz (2005), Gasieniec et al. (2005), Kowalski and Pelc (2007) and Huang et al. (2007b). We discuss broadcast routing further in Sect. 4.2.

3.5.2 CSMA-CA Based Scheduling Protocols

Several research approaches try to modify CSMA-CA based MAC to make it suitable to multi-hop mesh networks. Such ideas hold practical importance because they can be implemented using existing available 802.11 systems. A proposed MAC named DCMA (Data-driven Cut-through Medium Access) (Acharya et al. 2006) allows a packet to be forwarded from the Network Interface Card (NIC) only using MPLS like label-based forwarding. Such forwarding does not require IP route lookup or any other assistance from the forwarder's CPU. Packet's next hop is decided based on the label in RTS/ACK packet and the MAC-label table lookup in NIC. Due to combined RTS/CTS mechanism and pipeline kind of MAC-forwarding, DCMA reduces the number of channel access attempts and end-to-end latency. The study of Lim et al. (2007) extends such the label switching based MAC design for multi-radio multi-channel WMNs. It shows that with link layer forwarding in cut-through MAC, it is possible to make channel reservations in advance for packet's next hop simultaneously while receiving them from previous hop on a different channel. It provides modified channel access/reservation mechanism for this label-switched forwarding similar to 802.11 DCF which can reduce the end-to-end delay in multi-hop communication.

The RTS/CTS mechanism of 802.11 is often disabled in WMNs because of their over-conservative nature. In such cases, hidden terminal and exposed terminal problems can increase MAC collisions. The study of Mittal and Belding (2006) first proposed measurement based technique to mitigate the exposed terminal problem and improve spatial reuse. In the first phase, interference estimation technique of Padhye et al. (2005) is extended for detecting all potential exposed terminal combinations. Such information is then propagated in the network. Special control messages (RTSS—Request to Send Simultaneously, CTSS—Clear to Send Simultaneously) are then used whenever such transmissions with probable exposed terminals are encountered. This improves the overall simultaneous transmissions but requires large overhead of message transfers in the initial learning phase. The study of Hur et al. (2007) proposes the use of location information to avoid exposed terminal problem in 802.11 MAC protocol which can lead to better spatial reuse in mesh. Similarly, Raman and Chebrolu (2005) and Huang et al. (2007a) outline a busy-tone based solution for avoiding hidden terminal problem without interfering with data signals.

Other issues of CSMA-CA like rate control, fairness and carrier sense are also addressed for multi-hop networks. The study of Kim et al. (2006a) studies effectiveness of 802.11, 802.11e and 802.11n MACs on multi-hop mesh with different rate adaptation mechanisms. The study of Yang and Vaidya (2006) proposes spatial back-off algorithm which controls transmission rate and carrier sense threshold for current transmission to allow more number of other concurrent transmissions resulting into better spatial reuse. 802.11 MAC can be inherently unfair when used in multi-hop environment. Max-min models for per-flow fair bandwidth assignment to prevent such unfair MAC performance are provided by Raniwala et al. (2007) and Tang et al. (2006).

3.5.3 Other Scheduling Protocols

A scheduling mechanism is inherently fair and efficient if every node tries to transmit data depending on its backlog queue length compared to other nodes. In the work of Marbach (2007), a *distributed buffer* based design is proposed for distributed scheduling mechanism in wireless multi-hop networks. Here, transmission probability of every node is proportional to backlog of queue at its local buffer. If the arrival rate at a node is higher (often relay nodes or gateways), it gets more chance to occupy the medium for transmission. Though theoretically this may result into better fairness and higher network-wide throughput, its implementation may require modifications like busy tone, knowledge of offered load etc.

Directional antennas pose new set of challenges for link scheduling because of their different characteristics. The study of Li et al. (2005) provides insights about scheduling algorithm design for mesh network with directional antennas. Such a scheduling can benefit from higher transmission range and better spatial separation due to directional antennas. On the other hand, it also requires dealing with probably higher interference range, deafness and different sort of hidden terminal issues. 2-phase (2P) (Raman and Chebrolu 2005) scheduling protocol is suggested for rural area mesh networks with long point-to-point links and nodes having multiple directional antennas. In 2P, when a node switches from the transmission phase to the reception phase, its neighbors switch from the reception phase to the transmission phase and vice versa. This allows multiple receptions and transmissions possible at every node with multiple directional links (not possible by default in CSMA/CA MAC) but requires the network topology graph to be bipartite (a graph is bipartite if its vertices can be divided into two disjoint sets which are also independent sets of the graph). If the graph is not bipartite, it can be divided into several bipartite subgraphs and each such subgraph can be then assigned orthogonal channel to it as described by Raman (2006) and Dutta et al. (2007). This way 2P protocol can be used to scheduled transmission in each subgraph and transmissions between multiple such subgraph do not interfere with each other due to intelligent channel assignment.

3.5.4 802.11s MAC

With increasing number of CSMA/CA extensions for mesh networking been developed, it quickly became apparent that native MAC protocol of 802.11 is not well suited in multi-hop case. To address this, 802.11s follows a deterministic access ideology. The corresponding coordination function is referred as Mesh Deterministic Access (MDA).

> MDA is built on the idea of that contention for the medium and actual medium access should be clearly separated.

To do this, MDA divides the network function time into DTIM (Delivery Traffic Indication Message) intervals. Further, the DTIM intervals are mesh network wide phenomenon. The DTIM intervals are separated by beacons also referred as DTIM beacons. The beacons are used to maintain neighborhood information, network association, and most importantly (a rather coarse) synchronization of their local clocks.

Every DTIM interval consists of a beacon at the beginning, followed by MDA slots (refer to Fig. 3.23). Each MDA slot is 32 µs long. Many such consecutive slots provide an opportunity for data exchange between nodes. The transmission opportunities are

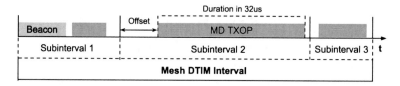

Fig. 3.23 Mesh Deterministic Access in IEEE 802.11s

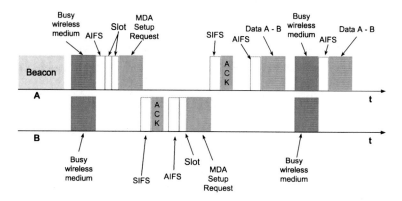

Fig. 3.24 MDA setup message

referred as MDAOP. For two mesh nodes to agree on communication, MDAOP protocol includes a handshake method. During the handshake, sender sends out an "MDA Setup Request". If the receiver agrees to communicate, it replies using "MDA Setup Reply". The MDA setup messages contain related information the subinterval that will be used for communication, relative offset within the subinterval and the total duration if the MDAOP data exchange. This is shown in Fig. 3.24. The receiving mesh node can either accept of decline to the MDA setup request. If declined, it can provide future opportunities in which it can communicate. The decision in made using the information regarding which other neighboring/interfering nodes are currently communicating, and how long the wireless medium will be occupied. If the mesh node agrees to setup an MDAOP, both mesh nodes broadcast a report containing the time of their data exchange. This record will be further used by other neighbors to plan their communications accordingly. Since there is an increased amount of information being exchanged in neighborhood, any such MDA data exchange is likely to be more successful as compared to a native CSMA/CA.

It is worth noting that at the start of MDAOP, there can be a collision since they do not follow a random back-off (as in Distributed Coordination Function—DCF of 802.11). To minimize such chances, MDA employs EDCA (Enhanced Distributed Channel Access) which enables prioritized medium access. In the case of MDA, the MDAOP owner (after the handshake) is given the highest opportunity for medium access. The EDCA mechanism is similar to that being used in 802.11e for providing QoS among various flows.

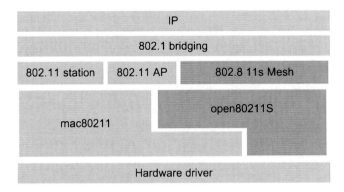

Fig. 3.25 Interoperation of 802.11s and other 802.11 in OPEN80211s

Fig. 3.26 General nature of 802.11s mesh network behavior

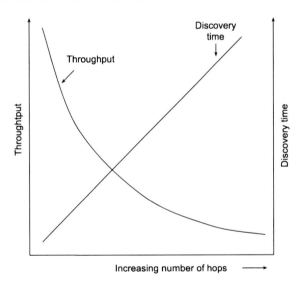

3.5.4.1 Performance Evaluation Using OPEN802.11S

Hiertz et al. (2010) has developed an open source implementation of 802.11s protocol that is compatible with Linux operating system. The Open80211s protocol stack is shown in Fig. 3.25. Using this implementation, Hiertz et al. (2010) showed the throughput and path discovery characteristics of 802.11s protocol. In the testbed setup, 12 mesh nodes are within in a single collision domain, and MAC filters are used in order to establish a chain-like multi-hop topology. One of the mesh nodes act as a gateway. As compared to native CSMA/CA used in 802.11, the throughput that can be achieved by a distant multi-hop node is still high. As Fig. 3.26 schematically shows, in fact increase in number of hops does reduce the overall throughput but the decrease is indeed not as sharp as observed in 802.11 in practice.

This shows that 802.11s can indeed offer performance gains, but there are various challenges in the draft that is being addressed continuously (Hiertz et al. 2010).

3.6 Channel/Radio Assignment

To mitigate the unavoidable consequences of interference, channel assignment mechanism tries to assign different non-interfering channels to the interfering links to increase the overall spatial reuse. The studies of Kyasanur et al. (2006) and Chereddi et al. (2006) discuss important design issues and practical challenges while designing multi-channel protocols for wireless mesh networks. As described by Subramanian et al. (2007), channel assignment protocols can be broadly classified in static, dynamic and hybrid schemes. We survey each class of channel assignment protocols next.

1. *Static channel assignment:* channel assignment remains unchanged over the course of network operation. Such mechanisms are often less adaptive to changing wireless conditions like external interference. On the other hand, such mechanisms are simpler and do not incur channel switching delays.
2. *Dynamic channel assignment:* channel assignment changes dynamically based on considerations like current interference, traffic demands, power allocation etc. This results into a more challenging design problem and also adds overhead of channel switching. Such mechanisms can be further classified into per link, per packet, per time-slot based mechanisms. As we discuss later, these channel assignment policies pose novel design problems like multi-channel hidden terminal, sporadic disconnections etc. but if carefully designed, they have the potential to achieve better system capacity.
3. *Hybrid channel assignment:* Some of the radios are assigned fixed channels while others switch their channels dynamically. These policies benefit from their partially dynamic design while inheriting simplicity of static mechanisms also.

3.6.1 Static Channel Assignment

Static channel assignment is a fixed assignment of channels to the radios of nodes which remains unchanged over the course of network operation. Such mechanisms are often less adaptive to changing wireless conditions like external interference and traffic. On the other hand, such mechanisms are simpler and do not incur channel switching delays.

In some of the earlier efforts to utilize multiple channels for network capacity enhancements, Adya et al. (2004) proposed a multi-radio unification protocol (MUP). MUP assigns different channels to different radios of a node and this assignment is identical for all nodes of the network. A node uses best quality channel out of its all radios for communicating with its neighbor. Though it improves performance with respect to the single channel assignment, number of channels utilized in the network is still restricted by the number of radios at nodes. Channel assignment problem can be modeled as edge-coloring of the network graph and related well-known heuristics or algorithms can be applied for the solution. Along the same lines, Marina and Das (2005) proposes a channel assignment algorithm CLICA (Connected Low Interference Channel Assignment) based on edge-coloring of the links in the connectivity graph. In the first phase of CLICA, every node greedily chooses colors for edges incident to it in a way such that the network connectivity is maintained. This choice is assisted by a weighted conflict graph so that the choice of link color minimizes the interference with conflicting links. Second phase handles multiple edges between the nodes and the unassigned radios at nodes which can be later utilized as per offered load using dynamic assignment.

Like link scheduling, conflict graph can also be used for channel assignment to incorporate interference relationship between the links. When dealing with multi-radio mesh nodes, the notion of conflict graph can be further extended to a per-radio case instead of per-node. The study of Subramanian et al. (2007) provides channel assignment algorithms for multi-radio WMNs with the objective of minimizing the co-channel interference while adhering to the interface assignment constraints. It uses conflict graph representation to capture the interference based conflicts between the links. This way, channel assignment problem of network graph turns out to be a vertex coloring problem in the corresponding conflict graph. Presented centralized algorithm tries to find such a coloring with condition that number of distinct channels assigned to the links incident to a node is no more than the number of interfaces available at the node. Distributed version of the algorithm tries to resolve the same using greedy heuristics of Max-K-cut problem (problem of assigning k colors to the vertices in such a way that the number of edges with endpoints of different colors is maximal). The efficiency of optimization algorithms are proved with semi-definite programming formulation. The study of Sen et al. (2007) shows that the link interference graphs (conflict graphs) belong to a special family of graphs called Overlapping Double-Disk (ODD) graphs. Such graphs can be created by having both endpoints of a link to possess a disk of radius half their interference range. If ODDs of two links intersect, it can be concluded that corresponding links interfere with each other. The channel assignment is performed by finding the independent sets using Polynomial-time Approximation Scheme (PTAS) in such ODD-based link interference graphs.

Changing the channel of a radio may cause several other nearby nodes to change the channels on their respective radios to maintain the symmetric links and channel dependencies. This is often referred as the *ripple effect* and it is an important design constraint addressed by Rad and Wong (2006). It proposes the design of logical topology from the actual physical topology while adhering to design constraints like channel dependency, ripple effect and hop count. Channel dependency constraint mentions that if multiple links are chosen in logical topology for the same radio at a node, all such links should be assigned the same channel. The choice of only a certain set of links out of the actual physical topology should also be carefully balanced to avoid long routing paths. Though there is no implicit consideration of interference, once all other constrains are formulated, actual radios are assigned channel based on the solution of logical topology. The study of Rad and Wong (2007) models the relation between channel assignment and radio assignment as binary vectors. Using link conflict graph for interference relationship, it models the achievable link rates as a function of these binary vectors. The joint problem is formulated as non-linear maximization problem, solution to which has been provided with two design schemes.

Similar to K-hop model of interference, Aryafar et al. (2008) present a novel edge coloring based channel assignment algorithm. The motivation is based on the observation that active links that are at distance of one hop from each other should be assigned different channel to avoid interference. This way channel assignment problem becomes Distance-1 edge coloring problem, which finds minimum number of colors such that any two active links at one hop distance are assigned different color. The problem being NP-complete, Aryafar et al. (2008) provide a heuristic for solution and describes a relevant MAC scheduling protocol based on the proposed solution.

Though majority of static channel assignment algorithms depend on graph coloring, there have been few other efforts also. The study of Vedantham et al. (2006) motivates the importance of component-based channel assignment in single-radio multi-channel ad-hoc networks. It proposes use of same channel for all links of a flow whenever multiple flows intersect at a node in the network. Different intersecting or contending flows may operate on different channels. Such design has merits of simplicity and lower switching delay. A combinatorial technique is used by Huang et al. (2006), named Balanced Incomplete Block Design (BIBD), for channel assignment. Specifically in BIBD, all nodes are assigned same number of distinct channels and each channel is assigned to same number of nodes. This way the network topology turns out to be a regular graph which has a good connectivity property. The algorithm presented by Huang et al. (2006) assigns channels such that certain connectivity is maintained and interference between same channel links are minimized.

Localized superimposed code based channel assignment algorithm is presented by Xing et al. (2007) where nodes use *channel code* (list of primary and secondary channels) to derive interference-free channel allocation. The approach taken by Zhu and Roy (2005) holds practical importance in terms of scalability and deployment where every node is equipped with two physical radios. It divides the mesh nodes into clusters, and cluster-head decides best intra-cluster channel to be used by detecting energy on every channel. Another radio at every node is dedicated for inter-cluster communication to handle the control and management messages. The study of Das et al. (2005) presents a ILP formulation for the channel assignment problem where the objective is to maximize total number of simultaneous transmissions on links while meeting the interference constrains. As one can see, all static policies discussed here can be used as a solution in network deployment and design phase but their inflexibility to adopt to changing conditions often require dynamic mechanisms of channel assignment which we discuss next.

3.6.2 Dynamic Channel Assignment

Such channel assignment changes dynamically based on considerations like current interference, traffic demands, power allocation etc. This results into a more challenging design problem and also adds overhead of channel switching. Such mechanisms can be further classified into per link, per packet, per time-slot based mechanisms. These channel assignment policies pose novel design problems like multi-channel hidden terminal, sporadic disconnections etc. but if carefully designed, they have the potential to achieve better system capacity.

Since every node in the network changes channels of its radios dynamically, nodes often require tighter coordination between them to avoid disconnections, deafness problems and multi-channel hidden terminal problem. Such issues make dynamic channel assignment mechanisms more and more complicated.

The multi-channel hidden terminal problem (So and Vaidya 2004) arises when channel selection is made during RTS/CTS exchange. When transmitter and receiver choose their channel for data transfer in RTS/CTS, it is possible that hidden terminal is listening on other channel. Such hidden terminal can never receive the choice of channel between sender and receiver, and may end up selecting same channel for its communication to some other node. This can result into collision at the receiver. To solve the problem of multi-channel hidden terminal, So and Vaidya (2004) proposed a multichannel MAC protocol (MMAC) which uses time synchronization between nodes in network just like 802.11 Power Saving Mode (PSM) using BECON intervals. In MMAC, in initial ATIM window all nodes tune to predefined control channel. All nodes having data to send, send ATIM message using control channel and also provides its preferred list of channels for data communication. Receivers choose a channel and sends back ATIM-ACK message. All other nodes hearing the channel choice choose their preferred channels different from it, avoiding the collision. After completion of ATIM window, actual data transfer takes place.

Sometimes it is not possible to dedicate a separate control channel due to lesser number of available orthogonal channels especially in standards like 802.11. Slotted Seeded Channel Hopping (SSCH) (Bahl et al. 2004) improves on MMAC by eliminating the need of such a control channel. In SSCH, each node switches channels in every slot based on its pseudo-random channel hopping schedule. Nodes have knowledge about other's channel hopping schedule. A sender wishing to send data designs its channel schedule in such a way that in some slot it achieves an overlap with receiver schedule. Such slotted design with switching channels can also benefit from the fact that distinct links can be active on different channels avoiding interference and increasing network performance by simultaneous communication. It is shown that such a random schedule can sometimes suffer from deafness problem (*missing receiver problem*) where transmitting node does not find intended receiver during the slot of communication.

Both SSCH and MMAC protocols require tight time synchronization between the nodes in the network. To avoid this problem, Maheshwari et al. (2006) proposes xRDT (Extended Receiver Driven Transmission) protocol which extends RDT (Shacham and King 1987), where sender switches to well-known fixed receiver channel for data transfer. xRDT uses additional busy tone interface to mitigate multichannel hidden terminal problem which can still happen in RDT. Proposed Local Coordination-based Multichannel MAC (Maheshwari et al. 2006) uses control and data window similar to MMAC (So and Vaidya 2004) without the need of global synchronization and busy tone interface. Senders use 802.11 based channel access mechanism in default channel to negotiate local schedules and channel usage during the control window.

CSMA-CA has been previously shown to be unfair even in single-cell infrastructure 802.11 networks. The study of Shi et al. (2006) first points out two fundamental coordination problem which causes flow starvation and unfairness in single-channel multi-hop CSMA networks—Information Asymmetry (IA) and Flow-in-the-middle (FIM) (Garetto et al. 2006). Multi-channel MAC can address these issues if designed carefully but it may itself can lead to problems like multi-channel hidden terminals or missing receiver problem (Garetto et al. 2006). Described Asynchronous Multi-channel Coordination Protocol (AMCP) uses one dedicated control channel. Nodes use the control channel to contend for preferred data channel using 802.11 DCF mechanism. Different from other previous protocols, in AMCP nodes can contend for data channels anytime without any specific synchronization. Selected data channel by sender-receiver is announced in RTS/CTS to other nodes which mark the channel to unavailable for that data communication time.

Several approaches rely on a central authority for performing the channel assignment and also try to accommodate real-time channel quality measurements. The study of Ramachandran et al. (2006) makes a significant contribution by developing a dynamic channel assignment algorithm which requires a centralized entity (gateway and channel assignment server). The proposed algorithm requires one radio at every mesh router to be dedicated for a common channel throughout the network. This is to maintain a connected back-bone topology, near optimal routing paths and non-interrupted flows of communication. It utilizes real-time measurements of all available channels to prioritize them based on their quality and effects of other co-located active networks (external sources of interference) on channel utilization. Based on this estimated co-channel interference, it develops a Multi-radio Conflict Graph (MCG). The MCG is build using a communication graph which represents every radio instead of every mesh node. This way, the number of assigned channels to a node is automatically restricted by the number of radios it has. Thus the MCG has one vertex for each possible radio-pair instantiating each possible link, and arcs representing conflicts between links. Once the MCG is created, gateway being central entity, initiates a breadth-first search for channel assignment based on the MCG and the information of channel priorities.

Because of its complexity in derivation and maintenance, only a few approaches have attempted to perform the channel assignment in distributed fashion. One of these (Shin et al. 2006) proposes a channel assignment heuristic (SAFE—Skeleton Assisted Partition Free) which assigns channels in a distributed fashion. With every node having K radios and N available channels in the network, if $N < 2K$ then every node randomly chooses K channels, leading to at least one common channel at every node. Nodes then communicate and choose different channels if their adjacent links have common channels. With $N > 2K$, SAFE finds a spanning subgraph of the network to maintain connectivity and assigns a default channel on it. As before, nodes choose a random set of channels and communicate with each other regarding their choices. Default channel is only used when other choices are not available without violating the interference limitations or the connectivity constraint.

All the approaches discussed to the point do not take traffic demands at nodes into consideration for the channel assignment. Most of the times, it is very difficult to derive a completely interference-free channel assignment solution. In such cases, if there exists a heuristic which can prioritize the links based on their importance, such a ranking of links can be helpful to perform channel assignment. The study of Rozner et al. (2007) motivated such need for traffic aware channel assignment in which partial or full information of current traffic is required. Such channel assignment ensures that nodes with high traffic demands are definitely assigned non-overlapping channels. Though presented algorithm is designed for WLANs, it can be applied to WMNs also for the high-traffic links near the gateway. Dynamic mechanisms are likely to incur higher overhead of control messages and are also more prone to ripple effect kind of real-time issues due to their fast adopting nature. The study by Gong and Midkiff (2005) uses routing control messages to propagate the information about channel assignment in K-hop neighborhood. It tries to assign non-conflicting channel to nodes during the Route Discovery and Reply processes itself to avoid any extra overhead. On the other hand, Kim and Shin (2007) define a framework for self-healing mesh network where network reconfigures itself minimally when faults like link failure occur. For example, in the case of high interference on a particular link, it forces minimal reconfiguration of the channel assignment and avoids network-wide ripple effects. Similar mechanism has also been proposed by Agrawal et al. (2006).

3.6.3 Hybrid Channel Assignment

In hybrid channel assignment schemes, some of the radios are assigned fixed channels while others switch their channels dynamically. These policies benefit from their partially dynamic design while inheriting simplicity of static mechanisms. As shown by Kyasanur and Vaidya (2005), in hybrid assignment all nodes try to assign different channel to their fixed radio. Node wishing to communicate switches its switchable radio to the channel of the fixed radio of the receiver.

IEEE 802.11 b/g standard provides 11 channels whose center frequencies are separated by 5 MHz and each channel spread around the center for 30 MHz. Though this results into only 3 non-overlapping channels, other partially overlapping channels can also be utilized for simultaneous communications if the interference caused between them is within a tolerable margin. Mishra et al. (2005a, b) presented first analytical reasoning about how partially overlapped channels can increase the spatial reuse. In WMNs, a node assigned a partially overlapping channel (POC) can help bridge the communication between nodes with entirely non-overlapping channels. The study of Mishra et al. (2005a, b) prove with examples that if designed carefully,

POC can provide routing flexibility as well as significant throughput enhancements. The study of Feng and Yang (2008) provided an evaluation of the usefulness of POC using testbed experiments and confirmed that when utilized carefully, POCs can improve network capacity by the factor of 2 in typical 802.11 b/g case. It provides a LP formulation for achievable network capacity in multi-hop networks using POCs. It also presents an interference model which captures the effects of partial interference of POCs. The interference range of POCs is much smaller than that of non-overlapping channels. This enables more simultaneous communications leading to a better spatial reuse as described by Mishra et al. (2005a, b) and Feng and Yang (2008).

With advancements in directional antennas and cognitive radio technologies, it is important that channel assignment mechanisms intelligently accommodate their characteristics. The study by Das et al. (2006a) uses directional antennas at every mesh router while designing mesh network. It incorporates the spatial separation provided by directional antennas in a channel assignment algorithm which improves on spatial reuse drastically. CogMesh (Chen et al. 2007) tries to address common control channel problem in cognitive radio based mesh network where spectrum access is dynamic. It tries to cluster the nodes on the basis of their detected spectrum hole and assigns it a control channel. With evolution of cognitive radio and software defined radio and their increasing usage in mesh, decisions of channel assignment and opportunistic access can become more and more complicated (Kyasanur et al. 2005). Because such adaptive radio technologies have capabilities to achieve true heterogeneity, their integration to mesh networks is imminent.

References

Acharya A, Ganu S, Misra A (2006) DCMA: a label switching MAC for efficient packet forwarding in multihop wireless networks. IEEE J Sel Areas Commun 24(11):1995–2004, doi:10.1109/JSAC. 2006.881636

Adya A, Bahl P, Padhye J, Wolman A, Zhou L (2004) A multi-radio unification protocol for IEEE 802.11 wireless networks. In: Broadband Networks, 2004. BroadNets 2004. Proceedings First International Conference on, pp 344–354. doi:10.1109/BROADNETS.2004.8

Agrawal D, Mishra A, Springborn K, Banerjee S, Ganguly S (2006) Dynamic interference adaptation for wireless mesh networks. In: Wireless mesh networks, 2006. WiMesh 2006. 2nd IEEE workshop on, pp 33–37. doi:10.1109/WIMESH.2006.288601

Aguayo D, Bicket J, Biswas S, Judd G, Morris R (2004) Link-level measurements from an 802.11b mesh network. SIGCOMM Comput Commun Rev 34(4):121–132. doi:10.1145/1030194.1015482

Akella A, Judd G, Seshan S, Steenkiste P (2007) Self-management in chaotic wireless deployments. Wirel Netw 13(6):737–755. doi:10.1007/s11276-006-9852-4

Akyildiz IF, Wang X, Wang W (2005) Wireless mesh networks: a survey. Comput Netw ISDN Syst 47(4):445–487. doi:10.1016/j.comnet.2004.12.001

Akyildiz IF, Lee WY, Vuran MC, Mohanty S (2006) Next generation/dynamic spectrum access/cognitive radio wireless networks: a survey. Comput Netw 50(13):2127–2159. doi:10.1016/j.comnet.2006.05.001

Alawieh B, Zhang Y, Assi C, Mouftah H (2009) Improving spatial reuse in multihop wireless networks—a survey. IEEE Commun Surv Tutor 11(3):71–91. doi:10.1109/SURV.2009.090306

Aryafar E, Gurewitz O, Knightly E (2008) Distance-1 constrained channel assignment in single radio wireless mesh networks. In: INFOCOM 2008. The 27th conference on computer communications. IEEE, pp 762–770. doi:10.1109/INFOCOM.2008.127

Bahl P, Chandra R, Dunagan J (2004) SSCH: slotted seeded channel hopping for capacity improvement in IEEE 802.11 ad-hoc wireless networks. In: MobiCom '04: proceedings of the 10th annual international conference on mobile computing and networking. ACM, New York, pp 216–230. doi:10.1145/1023720.1023742

Behzad A, Rubin I (2003) On the performance of graph-based scheduling algorithms for packet radio networks. IEEE Global Telecommun Conf (GLOBECOM '03) 6:3432–3436. doi:10.1109/GLOCOM.2003.1258872

Behzad A, Rubin I (2005) Impact of power control on the performance of ad hoc wireless networks. In: INFOCOM 2005. 24th annual joint conference of the IEEE computer and communications societies. Proceedings IEEE, vol 1, pp 102–113. doi:10.1109/INFCOM.2005.1497883

Björklund P, Värbrand P, Yuan D (2004) A column generation method for spatial TDMA scheduling in ad hoc networks. Ad Hoc Netw 2:405–418

Blough DM, Leoncini M, Resta G, Santi P (2005) Topology control with better radio models: implications for energy and multi-hop interference. In: MSWiM '05: proceedings of the 8th ACM international symposium on modeling, analysis and simulation of wireless and mobile systems. ACM, New York, pp 260–268. doi:10.1145/1089444.1089491

Brar G, Blough DM, Santi P (2006) Computationally efficient scheduling with the physical interference model for throughput improvement in wireless mesh networks. In: MobiCom '06: proceedings of the 12th annual international conference on mobile computing and networking. ACM, New York, pp 2–13. doi:10.1145/1161089.1161092

Bruno R, Conti M, Gregori E (2005) Mesh networks: commodity multihop ad hoc networks. IEEE Commun Mag 43(3):123–131. doi:10.1109/MCOM.2005.1404606

Burkhart M, von Rickenbach P, Wattenhofer R, Zollinger A (2004) Does topology control reduce interference? In: Levine N et al (eds) MobiHoc '04: proceedings of the 5th ACM international symposium on mobile ad hoc networking and computing. ACM, New York, pp 9–19. doi:10.1145/989459.989462

Chaporkar P, Kar K, Sarkar S (2005) Throughput guarantees through maximal scheduling in wireless networks. In: Proceedings of 43d annual allerton conference on communication, control and computing, pp 28–30

Chen T, Zhang H, Maggio G, Chlamtac I (2007) Topology management in CogMesh: a cluster-based cognitive radio mesh network. In: Communications, 2007. ICC '07. IEEE international conference on, pp 6516–6521. doi:10.1109/ICC.2007.1078

Chereddi C, Kyasanur P, Vaidya NH (2006) Design and implementation of a multi-channel multi-interface network. In: REALMAN '06: proceedings of the 2nd international workshop on multi-hop ad hoc networks: from theory to reality. ACM, New York, pp 23–30. doi:10.1145/1132983.1132988

Cicalese F, Manne F, Xin Q (2006) Faster centralized communication in radio networks. In: ISAAC, pp 339–348

Das A, Alazemi H, Vijayakumar R, Roy S (2005) Optimization models for fixed channel assignment in wireless mesh networks with multiple radios. In: Sensor and ad hoc communications and networks, 2005. IEEE SECON 2005. 2005 second annual IEEE communications society conference on, pp 463–474

Das S, Pucha H, Koutsonikolas D, Hu Y, Peroulis D (2006a) DMesh: incorporating practical directional antennas in multichannel wireless mesh networks. IEEE J Sel Areas Commun 24(11):2028–2039. doi:10.1109/JSAC.2006.881631

Das SM, Koutsonikolas D, Hu YC, Peroulis D (2006b) Characterizing multi-way interference in wireless mesh networks. In: WiNTECH '06: proceedings of the 1st international workshop on wireless network testbeds, experimental evaluation and characterization. ACM, New York, pp 57–64. doi:10.1145/1160987.1160999

Dimakis A, Walrand J (2006) Sufficient conditions for stability of longest-queue-first scheduling: second-order properties using fluid limits. Adv Appl Prob 38:505–521

Djukic P, Valaee S (2007a) Distributed link scheduling for TDMA mesh networks. In: Communications, 2007. ICC '07. IEEE international conference on, pp 3823–3828. doi: 10.1109/ICC.2007.630

Djukic P, Valaee S (2007b) Link scheduling for minimum delay in spatial re-use TDMA. In: INFOCOM 2007. 26th IEEE international conference on computer communications. IEEE, pp 28–36. doi:10.1109/INFCOM.2007.12

Draves R, Padhye J, Zill B (2004) Routing in multi-radio, multi-hop wireless mesh networks. In: MobiCom '04: proceedings of the 10th annual international conference on mobile computing and networking. ACM, New York, pp 114–128. doi:10.1145/1023720.1023732

Dutta P, Jaiswal S, Rastogi R (2007) Routing and channel allocation in rural wireless mesh networks. In: INFOCOM 2007. 26th IEEE international conference on computer communications. IEEE, pp 598–606. doi:10.1109/INFCOM.2007.76

Elkin M, Kortsarz G (2005) Improved broadcast schedule for radio networks. In: Symposium on discrete algorithms (SODA), pp 222–231

Feng Z, Yang Y (2008) How much improvement can we get from partially overlapped channels? In: Wireless communications and networking conference, 2008. WCNC 2008. IEEE, pp 2957–2962. doi:10.1109/WCNC.2008.517

Fuemmeler JA, Vaidya NH, Veeravalli VV (2006) Selecting transmit powers and carrier sense thresholds in CSMA protocols for wireless ad hoc networks. In: WICON '06: proceedings of the 2nd annual international workshop on wireless internet. ACM, New York, p 15. doi:10.1145/1234161.1234176

Gandham S, Dawande M, Prakash R (2008) Link scheduling in wireless sensor networks: distributed edge-coloring revisited. J Parallel Distrib Comput 68(8):1122–1134. doi: 10.1016/j.jpdc.2007.12.006

Gandhi R, Parthasarathy S, Mishra A (2003) Minimizing broadcast latency and redundancy in ad hoc networks. In: MobiHoc '03: proceedings of the 4th ACM international symposium on mobile ad hoc networking and computing. ACM, New York, pp 222–232. doi:10.1145/778415.778442

Gao J, Guibas LJ, Hershberger J, Zhang L, Zhu A (2001) Geometric spanner for routing in mobile networks. In: Proceedings of the 2nd ACM international symposium on mobile ad hoc networking and computing. MobiHoc '01. ACM, New York, pp 45–55. doi:10.1145/501422.501424,10.1145/501422.501424

Garetto M, Salonidis T, Knightly EW (2006) Modeling per-flow throughput and capturing starvation in CSMA multi-hop wireless networks. In: INFOCOM 2006. 25th IEEE international conference on computer communications. Proceedings, pp 1–13. doi:10.1109/INFOCOM.2006.194

Gasieniec L, Peleg D, Xin Q (2005) Faster communication in known topology radio networks. In: PODC '05: proceedings of the twenty-fourth annual ACM symposium on principles of distributed computing. ACM, New York, pp 129–137. doi:10.1145/1073814.1073840

Gomez J, Campbell A (2004) A case for variable-range transmission power control in wireless multihop networks. In: INFOCOM 2004. Twenty-third annual joint conference of the IEEE computer and communications societies, vol 2, pp 1425–1436

Gong M, Midkiff S (2005) Distributed channel assignment protocols: a cross-layer approach [wireless ad hoc networks]. IEEE Wireless Commun Netw Conf 4:2195–2200. doi:10.1109/WCNC.2005.1424857

Goussevskaia O, Oswald YA, Wattenhofer R (2007) Complexity in geometric SINR. In: MobiHoc '07: proceedings of the 8th ACM international symposium on mobile ad hoc networking and computing. ACM, New York, pp 100–109. doi:10.1145/1288107.1288122

Grönkvist J (1998) Traffic controlled spatial reuse TDMA in multi-hop radio networks. In: Personal, indoor and mobile radio communications, 1998. The ninth IEEE international symposium on, vol 3, pp 1203–1207. doi:10.1109/PIMRC.1998.731370

Grönkvist J, Hansson A (2001) Comparison between graph-based and interference-based STDMA scheduling. In: MobiHoc '01: proceedings of the 2nd ACM international symposium on mobile ad hoc networking and computing. ACM, New York, pp 255–258

Grönkvist J, Nilsson J, Yuan D (2004) Throughput of optimal spatial reuse TDMA for wireless ad-hoc networks. In: Vehicular technology conference, 2004. VTC 2004-Spring 2004. IEEE, 59th, vol 4, pp 2156–2160. doi:10.1109/VETECS.2004.1390655

Guo S, Yang OWW (2007) Energy-aware multicasting in wireless ad hoc networks: a survey and discussion. Comput Commun 30(9):2129–2148. doi:10.1016/j.comcom.2007.04.006

Gupta A, Lin X, Srikant R (2007) Low-complexity distributed scheduling algorithms for wireless networks. In: INFOCOM 2007. 26th IEEE international conference on computer communications. IEEE, pp 1631–1639. doi:10.1109/INFCOM.2007.191

Gupta P, Kumar P (2000) The capacity of wireless networks. IEEE Trans Inf Theory 46(2):388–404. doi:10.1109/18.825799.

Hajek B, Sasaki G (1988) Link scheduling in polynomial time. IEEE Trans Inf Theory 34(5):910–917. doi:10.1109/18.21215

Hamida EB, Chelius G, Fleury E (2006) Revisiting neighbor discovery with interferences consideration. In: PE-WASUN '06: proceedings of the 3rd ACM international workshop on performance evaluation of wireless ad hoc, sensor and ubiquitous networks. ACM, New York, pp 74–81. doi:10.1145/1163610.1163623

Hanzo L, Tafazolli R (2009) Admission control schemes for 802.11-based multi-hop mobile ad hoc networks: a survey. IEEE Commun Surv Tutor 11(4):78–108. doi:10.1109/SURV.2009.090406

Hiertz G, Denteneer D, Max S, Taori R, Cardona J, Berlemann L, Walke B (2010) IEEE 802.11s: the wlan mesh standard. IEEE Wireless Commun 17(1):104–111. doi:10.1109/MWC.2010.5416357

Hu YC, Perrig A (2004) A survey of secure wireless ad hoc routing. IEEE Secur Priv 2(3):28–39. doi:10.1109/MSP.2004.1

Huang HJ, Cao XL, Jia XH, Wang XL (2006) A BIBD-based channal assignment algorithm for multi-radio wireless mesh networks. In: Machine learning and cybernetics, 2006. International conference on, pp 4419–4424. doi:10.1109/ICMLC.2006.259095

Huang F, Yang Y, Zhang X (2007a) Receiver sense multiple access protocol for wireless mesh access networks. In: Communications, 2007. ICC '07. IEEE international conference on, pp 3764–3769. doi:10.1109/ICC.2007.620

Huang SH, Wan PJ, Jia X, Du H, Shang W (2007b) Minimum-latency broadcast scheduling in wireless ad hoc networks. In: INFOCOM 2007. 26th IEEE international conference on computer communications. IEEE, pp 733–739. doi:10.1109/INFCOM.2007.91

Hui KH, Lau WC, Yue OC (2007) Characterizing and exploiting partial interference in wireless mesh networks. In: Communications, 2007. ICC '07. IEEE international conference on, pp 102–108. doi:10.1109/ICC.2007.26

Hur SM, Mao S, Hou Y, Nam K, Reed J (2007) A location-assisted MAC protocol for multi-hop wireless networks. in: Wireless communications and networking conference, 2007. WCNC 2007. IEEE, pp 322–327. doi:10.1109/WCNC.2007.65

Illian J, Penttinen A, Stoyan H, Stoyan D (2008) Statistical analysis and modelling of spatial point patterns. Wiley-Interscience, New York

Iyer A, Rosenberg C, Karnik A (2006) What is the right model for wireless channel interference? In: QShine '06: proceedings of the 3rd international conference on quality of service in heterogeneous wired/wireless networks. ACM, New York, p 2. doi:10.1145/1185373.1185376

Jain K, Padhye J, Padmanabhan VN, Qiu L (2003) Impact of interference on multi-hop wireless network performance. In: MobiCom '03: proceedings of the 9th annual international conference on mobile computing and networking. ACM, New York, pp 66–80. doi:10.1145/938985.938993

Jia X, Kim D, Makki S, Wan PJ, Yi CW (2005) Power assignment for k-connectivity in wireless ad hoc networks. in: INFOCOM 2005. 24th annual joint conference of the IEEE computer and communications societies. Proceedings IEEE, vol 3, pp 2206–2211. doi:10.1109/INFCOM.2005.1498495

Jones CE, Sivalingam KM, Agrawal P, Chen JC (2001) A survey of energy efficient network protocols for wireless networks. Wirel Netw 7(4):343–358. doi:10.1023/A:1016627727877

Joo C, Lin X, Shroff N (2008) Understanding the capacity region of the greedy maximal scheduling algorithm in multi-hop wireless networks. In: INFOCOM 2008. The 27th conference on computer communications. IEEE, pp 1103–1111. doi:10.1109/INFOCOM.2008.165

Jun J, Sichitiu M (2003) The nominal capacity of wireless mesh networks. IEEE Wireless Commun 10(5):8–14. doi:10.1109/MWC.2003.1241089

Junhai L, Danxia Y, Liu X, Mingyu F (2009) A survey of multicast routing protocols for mobile ad-hoc networks. IEEE Commun Surv Tutor 11(1):78–91. doi:10.1109/SURV.2009.090107

Kashyap A, Ganguly S, Das S (2006) A measurement-based model for estimating transmission capacity in a wireless mesh network. In: WiNTECH '06: proceedings of the 1st international workshop on wireless network testbeds, experimental evaluation and characterization. ACM, New York, pp 103–104. doi:10.1145/1160987.1161012

Kawadia V, Kumar P (2003) Power control and clustering in ad hoc networks. In: INFOCOM 2003. Twenty-second annual joint conference of the IEEE computer and communications societies. IEEE, vol 1, pp 459–469. doi:10.1109/INFCOM.2003.1208697

Kawadia V, Kumar P (2005) Principles and protocols for power control in wireless ad hoc networks. IEEE J Sel Areas Commun 23(1):76–88. doi:10.1109/JSAC.2004.837354(410)23

Khalaf R, Rubin I (2004) Enhancing the throughput-delay performance of IEEE 802.11 based networks through direct transmissions. In: Vehicular technology conference, 2004. VTC2004-Fall 2004. IEEE 60th, vol 4, pp 2912–2916. doi:10.1109/VETECF.2004.1400593

Khan M, Kumar VA, Marathe M, Pandurangan G, Ravi S (2009) Bi-criteria approximation algorithms for power-efficient and low-interference topology control in unreliable ad hoc networks. In: INFOCOM 2009, IEEE

Kim KH, Shin KG (2006) On accurate measurement of link quality in multi-hop wireless mesh networks. In: MobiCom '06: proceedings of the 12th annual international conference on mobile computing and networking. ACM, New York, pp 38–49. doi:10.1145/1161089.1161095

Kim KH, Shin KG (2007) Self-healing multi-radio wireless mesh networks. In: MobiCom '07: proceedings of the 13th annual ACM international conference on mobile computing and networking. ACM, New York, pp 326–329. doi:10.1145/1287853.1287896

Kim S, Lee SJ, Choi S (2006a) The impact of IEEE 802.11 MAC strategies on multi-hop wireless mesh networks. In: Wireless mesh networks, 2006. WiMesh 2006. 2nd IEEE workshop on, pp 38–47. doi:10.1109/WIMESH.2006.288619

Kim TS, Lim H, Hou JC (2006b) Improving spatial reuse through tuning transmit power, carrier sense threshold, and data rate in multihop wireless networks. In: MobiCom '06: proceedings of the 12th annual international conference on mobile computing and networking. ACM, New York, pp 366–377. doi:10.1145/1161089.1161131

Kotz D, Newport C, Gray RS, Liu J, Yuan Y, Elliott C (2004) Experimental evaluation of wireless simulation assumptions. In: MSWiM '04: proceedings of the 7th ACM international symposium on modeling, analysis and simulation of wireless and mobile systems. ACM, New York, pp 78–82. doi:10.1145/1023663.1023679

Kowalski DR, Pelc A (2007) Optimal deterministic broadcasting in known topology radio networks. Distrib Comput 19(3):185–195. doi:10.1007/s00446-006-0007-8

Kumar S, Raghavan VS, Deng J (2006a) Medium access control protocols for ad-hoc wireless networks: a survey. Ad Hoc Netw 4(3):326–358

Kumar U, Gupta H, Das S (2006b) A topology control approach to using directional antennas in wireless mesh networks. in: Communications, 2006. ICC '06. IEEE international conference on, vol 9, pp 4083–4088. doi:10.1109/ICC.2006.255720

Kwon S, Shroff N (2007) Paradox of shortest path routing for large multi-hop wireless networks. In: INFOCOM 2007, pp 1001–1009. doi:10.1109/INFOCOM.2007.121

Kyasanur P, Vaidya N (2005) Routing and interface assignment in multi-channel multi-interface wireless networks. In: Wireless communications and networking conference, 2005, IEEE, vol 4, pp 2051–2056. doi:10.1109/WCNC.2005.1424834

Kyasanur P, Yang X, Vaidya NH (2005) Mesh networking protocols to exploit physical layer capabilities. In: IEEE workshop on wireless mesh networks, WiMesh

Kyasanur P, So J, Chereddi C, Vaidya N (2006) Multichannel mesh networks: challenges and protocols. IEEE Wireless Commun 13(2):30–36. doi:10.1109/MWC.2006.1632478

Li G, Yang LL, Corner WS (2005) Opportunities and challenges in mesh networks using directional antennas. In: Proceedings of IEEE workshop on wireless mesh networks, WiMesh

Li L, Halpern JY, Bahl P, Wang YM, Wattenhofer R (2001) Analysis of a cone-based distributed topology control algorithm for wireless multi-hop networks. In: PODC '01: proceedings of the twentieth annual ACM symposium on principles of distributed computing. ACM, New York, pp 264–273. doi:10.1145/383962.384043

Li L, Halpern JY, Bahl P, Wang YM, Wattenhofer R (2005) A cone-based distributed topology-control algorithm for wireless multi-hop networks. IEEE/ACM Trans Netw 13(1):147–159. doi:10.1109/TNET.2004.842229

Li N, Hou J, Sha L (2003a) Design and analysis of an MST-based topology control algorithm. In: INFOCOM 2003. Twenty-second annual joint conference of the IEEE computer and communications societies, IEEE, vol 3, pp 1702–1712

Li XY, Calinescu G, Wan PJ, Wang Y (2003b) Localized delaunay triangulation with application in ad hoc wireless networks. IEEE Trans Parallel Distrib Syst 14(10):1035–1047. doi:10.1109/TPDS.2003.1239871

Li XY, Wan PJ, Wang Y, Yi CW (2003c) Fault tolerant deployment and topology control in wireless networks. In: MobiHoc '03: proceedings of the 4th ACM international symposium on mobile ad hoc networking and computing. ACM, New York, pp 117–128. doi:10.1145/778415.778431

Li XY, Stojmenovic I, Wang Y (2004) Partial delaunay triangulation and degree limited localized bluetooth scatternet formation. IEEE Trans Parallel Distrib Syst 15(4):350–361. doi:10.1109/TPDS.2004.1271184

Li X, Mitton N, Simplot-Ryl I, Simplot-Ryl D (2011) A novel family of geometric planar graphs for wireless ad hoc networks. In: INFOCOM, 2011 proceedings IEEE, pp 1934–1942. doi:10.1109/INFCOM.2011.5934997

Lim JG, Chou CT, Nyandoro A, Jha S (2007) A cut-through MAC for multiple interface, multiple channel wireless mesh networks. In: Wireless communications and networking conference, 2007. WCNC 2007. IEEE, pp 2373–2378. doi:10.1109/WCNC.2007.443

Lin L, Lin X, Shroff N (2007) Low-complexity and distributed energy minimization in multi-hop wireless networks. In: INFOCOM 2007. 26th IEEE international conference on computer communications. IEEE, pp 1685–1693. doi:10.1109/INFCOM.2007.197

Lin X, Rasool S (2007) A distributed joint channel-assignment, scheduling and routing algorithm for multi-channel ad-hoc wireless networks. In: INFOCOM 2007. 26th IEEE international conference on computer communications. IEEE, pp 1118–1126. doi:10.1109/INFCOM.2007.134

Lin X, Shroff N (2006) The impact of imperfect scheduling on cross-layer congestion control in wireless networks. IEEE/ACM Trans Netw 14(2):302–315. doi:10.1109/TNET.2006.872546

Lin X, Shroff N, Srikant R (2006) A tutorial on cross-layer optimization in wireless networks. IEEE J Sel Areas Commun 24(8):1452–1463. doi:10.1109/JSAC.2006.879351

Ma H, Shin S, Roy S (2007) Optimizing throughput with carrier sensing adaptation for IEEE 802.11 mesh networks based on loss differentiation. In: Proceedings of IEEE. ICC 2007

Maheshwari R, Gupta H, Das S (2006) Multichannel MAC protocols for wireless networks. In: Sensor and ad hoc communications and networks, 2006. SECON '06. 2006 3rd annual IEEE communications society on, vol 2, pp 393–401. doi:10.1109/SAHCN.2006.288495

Marbach P (2007) Distributed scheduling and active queue management in wireless networks. In: INFOCOM 2007. 26th IEEE international conference on computer communications. IEEE, pp 2321–2325. doi:10.1109/INFCOM.2007.273

Marina M, Das S (2005) A topology control approach for utilizing multiple channels in multi-radio wireless mesh networks. In: Broadband networks, 2005. 2nd international conference on, vol 1, pp 381–390. doi:10.1109/ICBN.2005.1589641

Mishra A, Rozner E, Banerjee S, Arbaugh W (2005a) Exploiting partially overlapping channels in wireless networks: turning a peril into an advantage. In: IMC '05: proceedings of the 5th ACM SIGCOMM conference on internet measurement. USENIX Association, Berkeley, pp 29–29

Mishra A, Rozner E, Banerjee S, Arbaugh W (2005b) Using partially overlapped channels in wireless meshes. In: IEEE workshop on wireless mesh networks, WiMesh

Mittal K, Belding E (2006) RTSS/CTSS: mitigation of exposed terminals in static 802.11-based mesh networks. In: Wireless mesh networks, 2006. WiMesh 2006. 2nd IEEE workshop on, pp 3–12. doi:10.1109/WIMESH.2006.288617

Moaveni-Nejad K, Li X (2005) Low-interference topology control for wireless ad hoc networks. Ad-hoc Sensor Netw (an Int J) 1(1–2):41–64

Monks JP, Bharghavan V, mei W Hwu W (2001) A power controlled multiple access protocol for wireless packet networks. In: INFOCOM 2001. 20th annual joint conference of the IEEE computer and communications societies. Proceedings IEEE, pp 219–228

Moscibroda T, Wattenhofer R (2005) Coloring unstructured radio networks. In: SPAA '05: proceedings of the seventeenth annual ACM symposium on parallelism in algorithms and architectures. ACM, New York, pp 39–48. doi:10.1145/1073970.1073977

Moscibroda T, Wattenhofer R, Weber Y (2006) Protocol design beyond graph-based models. In: Proceedings of the 5th ACM SIGCOMM workshop on hot topics in networks (HotNets)

Muqattash A, Krunz M (2003) Power controlled dual channel (PCDC) medium access protocol for wireless ad hoc networks. In: INFOCOM 2003. Twenty-second annual joint conference of the IEEE computer and communications societies. IEEE, vol 1, pp 470–480. doi:10.1109/INFCOM.2003.1208698

Muqattash A, Krunz M (2005) POWMAC: a single-channel power-control protocol for throughput enhancement in wireless ad hoc networks. IEEE J Sel Areas Commun 23(5):1067–1084. doi:10.1109/JSAC.2005.845422

Nandiraju N, Nandiraju D, Santhanam L, He B, Wang J, Agrawal D (2007) Wireless mesh networks: current challenges and future directions of web-in-the-sky. IEEE Wireless Commun 14(4):79–89. doi:10.1109/MWC.2007.4300987

Narayanaswamy S, Kawadia V, Sreenivas RS, Kumar PR (2002) The COMPOW protocol for power control in ad hoc networks: theory, architecture, algorithm, implementation, and experimentation. In: European wireless conference

Newman MEJ, Watts DJ, Strogatz SH (2002) Random graph models of social networks. Proc Natl Acad Sc U S A 99:2566–2572

Padhye J, Agarwal S, Padmanabhan VN, Qiu L, Rao A, Zill B (2005) Estimation of link interference in static multi-hop wireless networks. In: IMC '05: proceedings of the 5th ACM SIGCOMM conference on internet measurement. USENIX Association, Berkeley, pp 28–28

Pantazis N, Vergados D (2007) A survey on power control issues in wireless sensor networks. IEEE Commun Surv Tutor 9(4):86–107. doi:10.1109/COMST.2007.4444752

Pathak P, Dutta R (2009) Impact of power control on capacity of large scale wireless mesh networks. In: IEEE ANTS 2009. doi:10.1109/ANTS.2009.5409865

Pathak P, Dutta R (2011) Impact of power control on capacity of tdm-scheduled wireless mesh networks. In: Communications (ICC), 2011 IEEE international conference on, pp 1–6. doi:10.1109/icc.2011.5962410

Pathak PH, Gupta D, Dutta R (2008) Loner links aware routing and scheduling wireless mesh networks. In: ANTS 2008. 2nd advanced networking and telecommunications system conference IEEE

Qiu L, Zhang Y, Wang F, Han MK, Mahajan R (2007) A general model of wireless interference. In: MobiCom '07: proceedings of the 13th annual ACM international conference on mobile computing and networking. ACM, New York, pp 171–182. doi:10.1145/1287853.1287874

Rad AHM, Wong VWS (2006) Wsn16-4: logical topology design and interface assignment for multichannel wireless mesh networks. In: Global telecommunications conference, 2006. GLOBECOM '06. IEEE, pp 1–6. doi:10.1109/GLOCOM.2006.985

Rad A, Wong V (2007) Joint channel allocation, interface assignment and MAC design for multi-channel wireless mesh networks. In: INFOCOM 2007. 26th IEEE international conference on computer communications. IEEE, pp 1469–1477. doi:10.1109/INFOCOM.2007.173

Ramachandran KN, Belding EM, Almeroth KC, Buddhikot MM (2006) Interference-aware channel assignment in multi-radio wireless mesh networks. In: INFOCOM 2006. 25th IEEE international conference on computer communications proceedings, pp 1–12. doi:10.1109/INFOCOM.2006.177

Raman B (2006) Channel allocation in 802.11-based mesh networks. in: INFOCOM 2006. 25th IEEE international conference on computer communications proceedings, pp 1–10. doi:10.1109/INFOCOM.2006.317

Raman B, Chebrolu K (2005) Design and evaluation of a new MAC protocol for long-distance 802.11 mesh networks. In: MobiCom '05: proceedings of the 11th annual international conference on mobile computing and networking. ACM, New York, pp 156–169. doi:10.1145/1080829.1080847

Ramanathan R, Rosales-Hain R (2000) Topology control of multihop wireless networks using transmit power adjustment. In: INFOCOM 2000. Nineteenth annual joint conference of the IEEE computer and communications societies proceedings IEEE, vol 2, pp 404–413. doi:10.1109/INFOCOM.2000.832213

Ramanathan S, Lloyd EL (1993) Scheduling algorithms for multihop radio networks. IEEE/ACM Trans Netw 1(2):166–177. doi:10.1109/90.222924

Ramaswami R, Parhi K (1989) Distributed scheduling of broadcasts in a radio network. In: INFO-COM '89. Proceedings of the eighth annual joint conference of the IEEE computer and communications societies technology: emerging or converging, IEEE, vol 2, pp 497–504. doi:10.1109/INFOCOM.1989.101493

Raniwala A, Pradipta D, Sharma S (2007) End-to-end flow fairness over IEEE 802.11-based wireless mesh networks. In: INFOCOM 2007. 26th IEEE international conference on computer communications IEEE, pp 2361–2365. doi:10.1109/INFOCOM.2007.281

Rappaport T (2001) Wireless communications: principles and practice, 2nd edn. Prentice Hall PTR, Upper Saddle River

Reis C, Mahajan R, Rodrig M, Wetherall D, Zahorjan J (2006) Measurement-based models of delivery and interference in static wireless networks. In: SIGCOMM '06: proceedings of the 2006 conference on applications, technologies, architectures, and protocols for computer communications. ACM, New York, pp 51–62. doi:10.1145/1159913.1159921

Rozner E, Mehta Y, Akella A, Qiu L (2007) Traffic-aware channel assignment in enterprise wireless lans. In: Network protocols, 2007. ICNP 2007. IEEE international conference on, pp 133–143. doi:10.1109/ICNP.2007.4375844

Santi P (2005) Topology control in wireless ad hoc and sensor networks. Wiley, New York

Santi P, Maheshwari R, Resta G, Das S, Blough DM (2009) Wireless link scheduling under a graded SINR interference model. In: FOWANC '09: proceedings of the 2nd ACM international workshop on foundations of wireless ad hoc and sensor networking and computing. ACM, New York, pp 3–12. doi:10.1145/1540343.1540346

Sen A, Murthy S, Ganguly S, Bhatnagar S (2007) An interference-aware channel assignment scheme for wireless mesh networks. In: Communications, 2007. ICC '07. IEEE international conference on, pp 3471–3476. doi:10.1109/ICC.2007.574

Shacham N, King P (1987) Architectures and performance of multichannel multihop packet radio networks. IEEE J Sel Areas Commun 5(6):1013–1025

Sharma G, Mazumdar RR, Shroff NB (2006) On the complexity of scheduling in wireless networks. In: MobiCom '06: proceedings of the 12th annual international conference on mobile computing and networking. ACM, New York, pp 227–238. doi:10.1145/1161089.1161116

Sharma G, Shroff N, Mazumdar R (2007) Joint congestion control and distributed scheduling for throughput guarantees in wireless networks. In: INFOCOM 2007. 26th IEEE international conference on computer communications. IEEE, pp 2072–2080. doi: 10.1109/INFOCOM.2007.240

Sharma S, Teneketzis D (2007) An externality-based decentralized optimal power allocation scheme for wireless mesh networks. In: Sensor, mesh and ad hoc communications and networks, 2007. SECON '07. 4th annual IEEE communications society conference on, pp 284–293. doi:10.1109/SAHCN.2007.4292840

Shi J, Salonidis T, Knightly EW (2006) Starvation mitigation through multi-channel coordination in CSMA multi-hop wireless networks. In: MobiHoc '06: proceedings of the 7th ACM international symposium on mobile ad hoc networking and computing. ACM, New York, pp 214–225. doi:10.1145/1132905.1132929

Shi Y, Hou YT, Liu J, Kompella S (2009) How to correctly use the protocol interference model for multi-hop wireless networks. In: MobiHoc '09: proceedings of the tenth ACM international symposium on mobile ad hoc networking and computing. ACM, New York, pp 239–248. doi:10.1145/1530748.1530782

Shin M, Lee S, ah Kim Y (2006) Distributed channel assignment for multi-radio wireless networks. In: Mobile adhoc and sensor systems (MASS), 2006. IEEE international conference on, pp 417–426. doi:10.1109/MOBHOC.2006.278582

So J, Vaidya NH (2004) Multi-channel MAC for ad hoc networks: handling multi-channel hidden terminals using a single transceiver. In: MobiHoc '04: proceedings of the 5th ACM international symposium on mobile ad hoc networking and computing. ACM, New York, pp 222–233. doi:10.1145/989459.989487

Stuedi P, Alonso G (2007) Log-normal shadowing meets sinr: a numerical study of capacity in wireless networks. In: Sensor, mesh and ad hoc communications and networks, 2007. SECON '07. 4th annual IEEE communications society conference on, pp 550–559. doi:10.1109/SAHCN.2007.4292867

Subramanian A, Gupta H, Das S (2007) Minimum interference channel assignment in multi-radio wireless mesh networks. In: Sensor, mesh and ad hoc communications and networks, 2007. SECON '07. 4th annual IEEE communications society conference on, pp 481–490. doi:10.1109/SAHCN.2007.4292860

Tang J, Xue G, Zhang W (2006) Maximum throughput and fair bandwidth allocation in multi-channel wireless mesh networks. In: INFOCOM 2006. 25th IEEE international conference on computer communications proceedings, pp 1–10. doi:10.1109/INFOCOM.2006.249

Tassiulas L, Ephremides A (1992) Stability properties of constrained queueing systems and scheduling policies for maximum throughput in multihop radio networks. IEEE Trans Autom Control 37(12):1936–1948. doi:10.1109/9.182479

Vargas E, Sayegh A, Todd T (2007) Shared infrastructure power saving for solar powered IEEE 802.11 WLAN mesh networks. In: Communications, 2007. ICC '07. IEEE international conference on, pp 3835–3840. doi:10.1109/ICC.2007.632

Vedantham R, Kakumanu S, Lakshmanan S, Sivakumar R (2006) Component based channel assignment in single radio, multi-channel ad hoc networks. In: MobiCom '06: proceedings of the 12th annual international conference on mobile computing and networking. ACM, New York, pp 378–389. doi:10.1145/1161089.1161132

von Rickenbach P, Schmid S, Wattenhofer R, Zollinger A (2005) A robust interference model for wireless ad-hoc networks. In: IPDPS '05: proceedings of the 19th IEEE international parallel and distributed processing symposium (IPDPS'05)—workshop 12. IEEE Computer Society, Washington, p 239.1. doi:10.1109/IPDPS.2005.65

Wattenhofer R, Li L, Bahl P, Wang YM (2001) Distributed topology control for power efficient operation in multihop wireless ad hoc networks. In: INFOCOM 2001. Twentieth annual joint conference of the IEEE computer and communications societies proceedings. IEEE, vol 3, pp 1388–1397. doi:10.1109/INFCOM.2001.916634

Xing K, Cheng X, Ma L, Liang Q (2007) Superimposed code based channel assignment in multi-radio multi-channel wireless mesh networks. In: MobiCom '07: proceedings of the 13th annual ACM international conference on mobile computing and networking. ACM, New York, pp 15–26. doi:10.1145/1287853.1287857

Xiong Y, Zhang Q, Wang F, Zhu W (2003) Power assignment for throughput enhancement (pate): a distributed topology control algorithm to improve throughput in mobile ad-hoc networks. In: Vehicular technology conference, 2003. VTC 2003-Fall 2003. IEEE 58th, vol 5, pp 3015–3019. doi:10.1109/VETECF.2003.1286177

Yang X, Vaidya N (2006) Spatial backoff contention resolution for wireless networks. In: Wireless mesh networks, 2006. WiMesh 2006. 2nd IEEE workshop on, pp 13–22. doi:10.1109/WIMESH. 2006.288600

Yang Y, Hou J, Kung LC (2007) Modeling the effect of transmit power and physical carrier sense in multi-hop wireless networks. In: INFOCOM 2007. 26th IEEE international conference on computer communications. IEEE, pp 2331–2335. doi:10.1109/INFCOM.2007.275

Yi S, Pei Y, Kalyanaraman S (2003) On the capacity improvement of ad hoc wireless networks using directional antennas. In: MobiHoc '03: proceedings of the 4th ACM international symposium on mobile ad hoc networking and computing. ACM, New York, pp 108–116. doi:10.1145/778415. 778429

Yucek T, Arslan H (2009) A survey of spectrum sensing algorithms for cognitive radio applications. IEEE Commun Surv Tutor 11(1):116–130. doi:10.1109/SURV.2009.090109

Zhang J, Jia X (2009) Capacity analysis of wireless mesh networks with omni or directional antennas. In: INFOCOM 2009. IEEE, pp 2881–2885. doi:10.1109/INFCOM.2009.5062251

Zhu J, Roy S (2005) 802.11 mesh networks with two-radio access points. In: Communications, 2005. ICC 2005. 2005 IEEE international conference on, vol 5, pp 3609–3615. doi:10.1109/ICC. 2005.1495090

Chapter 4
Mesh Design: Network Issues

4.1 Network-Level Design Challenges

Some research areas that the literature has addressed in the past several years deal with issues that only emerge, or become meaningful, in the context of the entire network. Routing is, of course, the quintessential such problem. In this chapter, we continue our examination of the traditional mesh design problem areas by focusing on such issues. Other such issues relate to the planning and deployment of the entire network—the placement of nodes, or optimization of placement costs. Modeling capacity and performance—of the network, not individual links or locales—also falls into this category. Rate control is another issue we discuss, since it is traditionally a mechanism in combating congestion, an emergent network phenomenon—made more critical by the reality of wireless interference. We also consider cognitive mesh design under this umbrella of network-wide issues.

As we mentioned before, the literature contains previous surveys of some of these topics, such as standard specific deployment issues (Bruno et al. 2005; Nandiraju et al. 2007), secure routing (Hu and Perrig 2004), multicast routing (Junhai et al. 2009), dynamic spectrum access (Akyildiz et al. 2006; Yucek and Arslan 2009). Table 4.1 is the companion to Table 3.1 summarizing our categorization of these areas. As before, they are intended purely as representative work, that may provide good starting points for the corresponding areas.

4.2 Routing

Just as in any other network, finding out high throughput routing paths is a fundamental problem in WMNs. Wireless mesh networks inherit many of their characteristics from traditional ad-hoc networks. Due to lesser consideration of mobility, increasing traffic demand and certain infrastructure-like design properties, routing protocols for WMNs have required exclusive focus from researchers. Table 3 presents a classifica-

P. H. Pathak and R. Dutta, *Designing for Network and Service Continuity in Wireless Mesh Networks*, Signals and Communication Technology, DOI: 10.1007/978-1-4614-4627-9_4, © Springer Science+Business Media New York 2013

Table 4.1 Categorization of WMN problems (network-wide issues)

Routing (Sect. 4.2)—Choosing routing paths to satisfy end-to-end traffic demands between nodes **Objective**: Low inter-path and intra-path interference, load balancing and hot-spot mitigation, higher reliability and throughput **Sample Literature**: Channel quality and diversity in multi-channel single-radio (So and Vaidya 2005) and multi-channel routing (Draves et al. 2004), opportunistic routing protocol (Biswas and Morris 2005), hot-spot analysis with straight line routing (Kwon and Shroff 2007)
Rate control and congestion control (Sect. 4.5)—TCP-like congestion control methods that work despite loss and bias introduced by wireless medium, including multihop **Objective**: Throughput, congestion control, fairness **Sample Literature**: general TCP survey (Lochert et al. 2007), addressing spatial bias (Mancuso et al. 2010), neighborhood awareness (Rangwala et al. 2008)
Network planning and deployment (Sect. 4.3)—Topological and deployment factors, gateway placement **Objective**: Network expansion in non-cooperative environment, load balancing with intelligent gateway placement **Sample Literature**: Study of deployment and topological factors (Robinson and Knightly 2007)
Performance modeling and capacity analysis (Sect. 4.4)—Understanding best and worst case theoretical capacity **Objective**: Performance analysis and estimation of system capacity and newly developed protocols **Sample Literature**: Best case theoretical throughput of WMNs (Gupta and Kumar 2000), Capacity of multi-channel WMNs (Kyasanur and Vaidya 2005b)

tion of WMN routing protocols and summarizes their characteristics and objectives. We next survey research in each of these individual categories one by one in the following subsections.

Routing metrics and protocols of wireless multi-hop networks differ from other traditional routing protocols due to dynamic and unpredictable nature of wireless medium. WMNs display relatively stable topological behavior due to lack of mobility but still underlying issues of link quality and the interference remain the same. This has motivated design and development of various new routing metrics and protocols for WMNs. First, we discuss some of the routing metrics that have been proposed for WMNs and then survey routing protocols which actually utilize them.

4.2.1 Routing Metrics for Wireless Mesh Networks

Naively utilizing the hop-count as routing metric in mesh has proven to be inefficient (Couto et al. 2003b) as it does not take dynamic characteristics of wireless medium such link quality, interference etc. into consideration. As mentioned by Yang and Kravets (2005), WMNs differ from other wireless ad-hoc networks in terms of their static nodes. Though inherent wireless medium is similar, the links between nodes are fairly stable and display relatively higher constant characteristics. These properties require exclusive routing metric and protocol design for WMNs. Considering the routing metrics first, Yang and Kravets (2005) provide detailed explanation of characteristics that a mesh routing metric should possess. It shows that the metric should provide stable, good performance (in terms of throughput or delay), computationally efficient and loop-free routing paths. Though it has been shown that topology-dependent routing metrics are more stable in relatively static environments like mesh, many recent metrics still consider dynamic wireless conditions.

Below, we present some of metrics proposed in literature for routing in WMNs. A more detailed comparison between a few of them can be found in the work by Yang and Kravets (2005).

1. ETX (Couto et al. 2003a): *Expected transmission count (ETX)* is the estimated number of transmissions (including retransmission) required to send a data packet over the link. In this terms, if link has a forward delivery ratio d_f (probability that data packet successfully arrives at receiver) and backward delivery ratio of d_r (probability that ACK is received by sender) then its ETX value can be defined as follows

$$ETX = \frac{1}{d_f \times d_r} \tag{4.1}$$

   $d_f \times d_r$ shows that packet is transmitted with success in forward direction and ACK is also successfully received in backward direction. The total ETX of a path is summation of ETX of all links on the path.

2. ETT (Draves et al. 2004): *Expected transmission time (ETT)* improves over ETX by considering bandwidth also while assigning metric to a link. If S is the size of the packet and B is bandwidth of the link then ETT can be defined as follows

$$ETT = ETX \times \frac{S}{B} \tag{4.2}$$

   This way, ETT of the link captures the time taken for successfully transmitting a packet on the link.

3. WCETT (Draves et al. 2004): *Weighted cumulative ETT* improves over ETT by considering the channel diversity along the path. As different links on paths might have different channels assigned to it, it is important to capture the effect of sum of transmission times of links on every channel. Let X_j be the sum of transmission times of links on channel j as follows

$$X_j = \sum_{\substack{link\ i\ is\ on\ channel\ j}} ETT_i \quad 1 \leq j \leq k \tag{4.3}$$

Now, WCETT can be defined as follows

$$WCETT = (1 - \beta) \times \sum_{i=1}^{n} ETT_i \; + \; \beta \times \max_{1 \leq j \leq k} X_j \tag{4.4}$$

Thus, WCETT finds routing paths with least ETT values and highest channel diversity. WCETT is proven to be non-isotonic (Yang and Kravets 2005) (a metric has isotonic property if it ensures that order of weights of two paths are preserved if they are appended or prefixed by a common third path) which requires very efficient algorithms to find minimum weight paths. The study of Korkmaz and Zhou (2006) discusses how to use iterative line search technique to efficiently find WCETT based optimal or near-optimal paths using Dijsktra's algorithm.

4. MIC (Yang et al. 2005, 2006): *Metric of interference and channel switching* improves over ETT by considering inter-flow and intra-flow interference using IRU (Interference-aware Resource Usage) and CSC (Channel Switching Cost) components of links. IRU of a link ij operating on channel c also includes its ETT and can be defined as below

$$IRU_{ij}(c) = ETT_{ij}(c) \times |N_i(c) \cup N_j(c)| \qquad (4.5)$$

$|N_i(c) \cup N_j(c)|$ is the number of neighboring nodes interfered due to activity of a link ij on channel c. To consider intra-flow interference, every node on the routing path is assigned CSC value to it. CSC of a node x is lesser if previous link where x was receiver and next link where x is sender are on different channels. CSC value us higher if both incoming and outgoing links are on the same channel as it introduces more intra-flow interference. MIC of a routing path p can be expressed as below

$$MIC(p) = \alpha \sum_{link\ ij \in p} IRU_{ij} + \sum_{node\ i \in p} CSC_i \qquad (4.6)$$

Here, $\alpha = \frac{1}{N \times min(ETT)}$ which tries to balance the load in the network.

5. MCR (Kyasanur and Vaidya 2006): *Multi-channel routing* metric improves over WCETT by considering switching costs required for channel switching on different links along the path. WCETT does not capture the effect of switching delay for links active on different channels on a path. MCR adds switching delay to metric so that switching delay at every link does not take away the benefits achieved from hybrid channel assignment (Kyasanur and Vaidya 2005a). Let $SC(c_i)$ be the switching cost of ith hop on a path, operating on channel c_i then MCR combines the effect of channel quality, diversity and switching delay as follows

$$MCR = (1 - \beta) \times \sum_{i=1}^{n} (ETT_i + SC(c_i)) \quad + \quad \beta \times \max_{1 \leq j \leq k} X_j \qquad (4.7)$$

6. WCCETT (Jiang et al. 2007): *Weighted Cumulative Consecutive ETT* also proposes a way to extend WCETT for better consideration of intra-flow interference. If we refer consecutive hops of a path which are operating on same channel as segment then WCCETT can defined as below

$$Y_j = \sum_{\text{link } i \text{ is on segment } j} ETT_i \qquad 1 \leq j \leq k \qquad (4.8)$$

$$WCCETT = (1 - \beta) \times \sum_{i=1}^{n} ETT_i \; + \; \beta \times \max_{1 \leq j \leq k} Y_j \qquad (4.9)$$

This way, WCCETT selects a path with more channel diversity (smaller segments) compared to WCETT yielding lesser intra-flow interference.

7. iAWARE (Subramanian et al. 2006): WCETT does not capture the inter-flow interference and may end up choosing congested routing paths. iAWARE uses physical interference model for calculating inter-flow interference. The study of Subramanian et al. (2006) defines *Interference Ratio (IR$_l$)* for link l from u to v where $IR_l = min(IR_i(u), IR_i(v))$ and $IR_i(v) = \frac{SINR_i(v)}{SNR_i(v)}$. iAWARE is defined as below

$$iAWARE_l = \frac{ETT_l}{IR_l} \qquad (4.10)$$

iAWARE of path is calculated in similar way as ETT.

8. ETOP (*Expected number of Transmissions On a Path*) (Jakllari et al. 2007, 2008): As mentioned by Jakllari et al. (2007) and (2008), ETX metric does not take into account that practically if certain number of link layer transmissions are unsuccessful, then packet is dropped and transport layer at the source node re-initiates the end-to-end transmission. In this case, if the lossy link is closer to the destination than source, most of the link transmissions from source to the lossy link are often wasted in unsuccessful end-to-end attempts. ETOP can be defined as expected number of transmissions required for delivering a packet over a path. ETOP takes into account the effect of relative position of links on path together with number of links and link quality.

9. METX (Roy et al. 2006): *Multicast ETX* (originally $C(s, d)$ (Dong et al. 2005)) captures the total expected number of transmissions required by all nodes along the source-destination path so that destination receives at least one packet successfully. It is formally defined as follows

$$METX = \sum_{l=1}^{n} \frac{1}{\prod_{i=l}^{n}(1 - Perr_i)} \qquad (4.11)$$

where l denotes lth link on n-hop path and $Perr_l$ is the error rate of the link.

10. SPP (Roy et al. 2006): *Success Probability Product* (originally EER in (Banerjee and Misra 2002)) is proposed for multicast routing in WMNs. It is similar to METX and can be defined as $SPP = \prod_{l=1}^{n} d_{fl}$ where $d_{fl} = 1 - Perr_l$. Considering link layer broadcast in multicast, SPP reflects the probability that destination receives the packet without error. Routing protocol should choose the path which has minimum $1/SPP$ (Table 4.2).

Table 4.2 Categorization of WMN routing strategies

Routing strategy	Objective, characteristics	Sample literature
MANET-Like Routing (Sect. 4.2.2)	Reactive or proactive, adapt MANET routing protocols to relatively stable and high bandwidth environment of WMNs, incorporate a WMN routing metric in existing protocol	Proactive: AODV-ST (citealt104), reactive: OLSR (Clausen and Jacquet 2003), B.A.T.M.A.N. (BATMAN 2007)
Opportunistic Routing (Sect. 4.2.3)	Hop-by-hop routing, exploit fortunate long distance receptions to make faster progress towards destination	Ex-OR (Biswas and Morris 2005)
Multi-Path Routing and Load Balancing (Sect. 4.2.4)	Maintain redundant routes to destination, determine divergent routes to mitigate the crowded center effect, load balancing and fault tolerance	Multi-path routing (Ganjali and Keshavarzian 2004; Pham and Perreau 2003), load balancing (Popa et al. 2007)
Geographic Routing (Sect. 4.2.5)	Utilize location information for forwarding in large mesh networks	Efficient geographic routing (Lee et al. 2005b)
Hierarchical Routing (Sect. 4.2.6)	Divide network into clusters and perform routing for better scalability	Clustering and hierarchical routing (Ramanathan and Steenstrup 1998)
Multi-Radio and Multi-Channel Routing (Sect. 4.2.7)	Accommodate intra-path and inter-path interference, consider channel assignment constraints and switching cost	Multichannel routing (Kyasanur and Vaidya 2006)
Multicasting Protocols (Sect. 4.2.8)	Adapt existing multicast mechanisms of ad-hoc networks to WMNs	Multicasting in WMNs (Roy et al. 2006)
Broadcast Routing (Sect. 4.2.9)	Minimum latency broadcasting with least number of retransmissions, adapting to multi-channel environment	Broadcasting in multi-channel WMNs (Qadir et al. 2006)

4.2.2 Traditional MANET-Like Routing Protocols

The MANET routing protocols were designed for mobile wireless nodes, intermittent links and frequently changing topologies. Such protocols often rely on flooding for route discovery and maintenance. Direct employment of such protocols is not suitable for relatively static mesh networks for various reasons.

Traditional MANET like protocols can be largely classified in reactive and proactive routing protocols. AODV (Perkins et al. 2003), DSR (Johnson et al. 2007) etc. are **reactive routing** protocols in which a route discovery is initiated only on demand from any source node. Links in WMNs are fairly stable over a longer period of time and likely to carry relatively stable backbone-like traffic. Flooding messages for on-demand route discovery can induce high unnecessary overhead in WMNs (Yang and Kravets 2005). Also, such protocols mostly use hop-count as routing metric which is not suitable for wireless medium because it can lead to shorter yet low throughput routing paths (Couto et al. 2003b; Yang and Kravets 2005).

Proactive routing protocols are table-driven protocols which require flooding in case of link failure and use hop-count as primary metric for routing. They do not take link quality or any other dynamic wireless characteristics like intermediate packet losses in consideration. Many of the proactive routing protocols have been adopted or specifically designed for WMNs. As an example, OLSR (Clausen and Jacquet 2003) has recently accommodated feature for link quality sensing and it is being adapted for mesh implementations. Similarly, Babel (2007) is also a proactive routing protocol based distance vector routing and utilizes link ETX values for maintaining better quality routes. Hop-by-hop forwarding (e.g. opportunistic routing) is better suited for mesh than table-driven routing protocols due to its simplicity and possible adaptation to link dynamics (Yang and Kravets 2005). B.A.T.M.A.N. (Better Approach To Mobile Ad-hoc Networking) routing protocol (BATMAN 2007) tries to adopt such forwarding ideology in which every node maintains logical *direction* towards the destination and accordingly chooses next-hop neighbor while routing. A useful empirical comparison of these proactive routing protocols can be found in Abolhasan et al. (2009).

Instead of developing new routing protocols for WMNs, many researchers have proposed modifications to the above mentioned MANET-like routing protocols. Most of such protocols try to adapt to the characteristics of WMNs such as lower mobility, stable routes etc. Also, variety of such protocols utilizes previously discussed routing metrics. Following are a few examples of such protocols:

- *AODV-ST* (Ramachandran et al. 2005): Ramachandran et al. (2005) provide AODV-ST (spanning tree) routing protocol which improves on AODV in several way to adapt to WMN characteristics. To avoid repetitive reactive route discovery with flooding, AODV-ST maintains spanning tree paths rooted at gateway from the nodes. It can incorporate high throughput metrics like ETT, ETX etc. for high performance spanning tree paths. AODV-ST also uses IP-IP encapsulation for avoiding large routing tables at relay nodes and can also perform load balancing for gateways.
- *AODV-MR* (Subramanian et al. 2006): Subramanian et al. (2006) present multi radio extension for AODV protocol where each node has multiple radios and channel assignment is performed with some pre-determined static technique. AODV-MR uses iAWARE metric with bellman-ford algorithm to find efficient low interference paths. Links on such paths display low intra-flow and inter-flow interference together with good link quality.
- *ETOP-R* (Jakllari et al. 2008, 2007): ETOP-R routing protocol uses ETOP routing metric described earlier for finding shortest path using Dijkstra's shortest path algorithm. Practically, ETOP-R has been implemented with modified source routing protocol DSR.
- *THU-OLSR (Timer-Hit-Use OLSR)* (Jiang et al. 2007): An interval optimization algorithm is presented by Jiang et al. (2007) which adaptively adjusts control message intervals of OLSR based on the mobility. The *hello* interval and *topology control* interval of OLSR are set based on neighbor's status and multi-point relay

(MPR) selector's status. This informed values of intervals are then utilized in THU-OLSR.

- *PROC* (Hu et al. 2007): In Progressive ROute Calculation (PROC) protocol, source node first establishes a preliminary route to destination using broadcast. Destination then initiates building of a minimum cost spanning tree to source with the nodes around the preliminary route. The source uses this optimal route for future data transfer.

4.2.3 Opportunistic Routing Protocols

As we discussed previously that traditional shortest path routing and traditional ad-hoc routing protocols may not be sufficient for mesh. Recently, opportunistic routing protocols have been proposed to exploit unpredictable nature of wireless medium. Unlike all previous approaches, opportunistic routing protocol defers the next hop selection after the packet has been transmitted. Meaning, if a packet fortunately makes it to a far distant node than expected, such useful transmissions should be fully exploited. Though there are many advantages of such mechanisms like faster progress towards the destination, it requires complex coordination between the transmitters regarding the progress of the packets. Many protocols have been developed based on such idea which we discuss below.

- *Ex-OR* (Biswas and Morris 2005): An important opportunistic routing protocol was proposed by Biswas and Morris (2005) which displayed its direct applicability in WMNs. In proposed routing protocol (called Ex-OR), sender broadcasts batch of packets with a list of potential forwarders in order of their chances to reach destination. The highest priority forwarder forwards the packets from its buffer each having copy of sender's estimate of highest priority node which should have received the packets. To avoid blind flooding, it maintains information about which packets have been received by the intermediate nodes. The packets which are not received and acknowledged by higher priority forwarders are forwarded by the other forwarders in the list. The process continues until the batch of packets reaches the destination.
- *SOAR* (Rozner et al. 2006): Simple Opportunistic Routing Protocol (SOAR) proposed by Rozner et al. (2006) improves on Ex-OR in certain ways and efficiently supports multiple flows in WMNs. First, it requires the nodes forwarding packets to be near the shortest path (least ETX) from source to destination to avoid packets being misdirected. Secondly, it adds a timer based low overhead distributed mechanism to coordinate between the forwarders regarding when and which packets to forward. Higher priority nodes having smaller timer values forwards first upon it expiration. Other forwarders listening to it, discards the redundant packets which avoid unnecessary flooding without any extra coordination overhead.
- *MORE* (Chachulski et al. 2007): Ex-OR requires high amount of coordination between the forwarders and inherently cannot take advantage of spatial reuse. MORE (Chachulski et al. 2007) (MAC-independent Opportunistic Routing and Encoding) extends the Ex-OR with network coding. Here, packets are randomly

mixed before forwarding to avoid the redundant packet transmissions without any need of special scheduling or coordination. Similar approaches are presented by Katti et al. (2006); Sengupta et al. (2007).

- *ROMER* (Yuan et al. 2005): Similarly, in Resilient Opportunistic Mesh Routing (ROMER) (Yuan et al. 2005) protocol, a packet traverses through the nodes only *around* long-term and stable minimum cost path. These nodes build a dynamic forwarding mini-mesh of nodes on the fly. In between, each intermediate node opportunistically selects transient high throughput links to take advantage of short-term channel variations. This way, ROMER deals with node failures and link losses, and also benefits from opportunistic high throughput routing.

4.2.4 Multi-Path Routing and Load Balancing

As mentioned by Nandiraju et al. (2006), using traditional routing approaches and metrics, many mesh routers may end up choosing already congested routing paths to reach the gateway nodes. This can lead to low performance due to highly loaded routing paths. The study of Nandiraju et al. (2006) proposes a routing protocol called MMESH (Multipath Mesh), in which every node derives multiple paths to reach gateway node using the source routing. It then performs load balancing by selecting one of the least loaded paths. A large set of multi-path routing protocols are reviewed by Tsai and Moors (2006).

Other multi-path routing mechanisms have been previously proposed by Lee and Gerla (2000) and Pham and Perreau (2003) for ad-hoc networks. Interestingly, Ganjali and Keshavarzian (2004) claim that unless and until very large number of paths (infeasible in practice) are used in multi-path routing, single path routing performs almost as good as multi-path routing. In such cases, more routes to destination do not help much in balancing the load throughout the network. This is in line with common belief of generation of hot-spots in multi-hop wireless networks. When shortest path or straight line routing is used, most of the routing paths pass through a certain region (center of network) creating a highly congested, security prone area. Nodes in such area have to relay disproportionate amount of traffic for other nodes and often suffer from severe unfairness. Recently, Kwon and Shroff (2007) showed that relay load on the network mainly depends on the offered traffic pattern. When shortest path routing is used with random traffic pattern, it can give rise to different load distribution, generating hot-spot at different places in the network.

Problem of modeling the relay load of nodes in network is been addressed by a few research efforts. In uniform topologies, relay load is often modeled as a function of node's distance from the center (Lassila 2006; Popa et al. 2007). Recently, relay load of a node has also been modeled probabilistically as a function of perimeter of node's Voronoi cell (Kwon and Shroff 2007). Though such modeling works in uniform topologies and traffic, relay load estimation in arbitrary topologies is still an open problem. Similarly, finding ways to evenly distribute the relay load in the network is also an open research issue and is being actively investigated. Current approached for relay load balancing depends mainly on transforming Euclidean network graph on symmetric spaces like sphere or torus which do not show such *crowded center* characteristics. Many *divergent, center-avoiding* routing mechanisms described below have been proposed by researchers to try and balance the relay load among the nodes.

- *Curve-ball Routing*: Popa et al. (2007) and Li and Wang (2008a) present an approach for load balancing in which stereographic projection is used to map the Euclidean node positions on a sphere. The routes between the source and destination are then found using great circle distance on sphere and then they are mapped back to the actual plane in network. Such routes often results in circular arc shaped forwarding which is claimed to be distributing the load in network since they intentionally avid passing through the center.
- *Outer-space Routing*: Mei and Stefa (2008) proposed the concept of routing in outer space in which original network space is mapped onto a symmetric outer space (torus). The shortest routing paths between nodes in such outer space will symmetrically distribute the relay load in the entire network. Such paths are then used for routing in the original network to avoid routing via hot-spots.
- *Manhattan routing*: Durocher et al. (2008) proposed a divergent routing scheme in which source forwards the packet to an intermediate node which is near the intersection of horizontal/vertical lines passing through the source and destination.

Similarly, several other similar load balancing mechanisms are described and analyzed by Esa (2000). As shown by Li and Wang (2008b) and Gao and Zhang (2009), such routing mechanisms display trade-off between stretch-factor of routing paths and actual load balancing.

4.2.5 Geographic Routing

The MANET routing protocols often assume the availability of location information at nodes to facilitate intelligent data forwarding. WMNs can benefit from such location information and several routing protocols are presented for such geographic routing and related issues. The study of Lee et al. (2005b) proposes an efficient geographic routing protocol where packets are forwarded towards the neighbor closest to the destination. Forwarding decisions are made on hop-by-hop basis. It proposes a link metric called Normalized Advance (NADV) which is defined as

$$NADV(n) = \frac{ADV(n)}{Cost(n)} \qquad Where \qquad ADV(n) = D(S) - D(n) \qquad (4.12)$$

Here, $D(x)$ denotes the distance from node x to destination and $cost(n)$ can be any cost factor like packet error rate, delay etc. This way NADV reflects the amount of progress made towards the destination per unit cost.

If non-uniform topologies, geographic forwarding may result into inefficiency if an intermediate node may not find any other node towards destination. Such regions are called *routing holes* by Subramanian et al. (2007), which proposes an oblivious routing scheme with fixed number of routing holes for random source destination pair traffic. Randomized routing which constructs random path around the hole is proposed by Subramanian et al. (2008) where arbitrary number of such holes are considered.

Extending the current state of art in location aware routing, Cheng et al. (2006) propose a rendezvous based routing which only requires local information about the relative direction of 1-hop neighbors at every node. Node wishing to transmit forwards the request to all four orthogonal directions and subsequent nodes forward the request in opposite direction (from which they received) until route to the destination is found. It is claimed that such routing mechanism is highly likely to find paths due to the fact that pair of orthogonal lines centered at two different points in the plane will intersect with a high probability. Similar approaches for geographic routing are presented for MANETs and sensor networks by Fang et al. (2005), Kuhn et al. (2003) and Tang et al. (2007).

Kwon and Shroff (2006) study the problem of energy-efficient interference-based routing with respect to new flow admission in multi-hop wireless networks. The problem is first formulated as energy minimization with bandwidth constraint. It is then converted in terms of SINR constraint and matrix arithmetic is used for solving it. For any scheduling mechanism, the proposed routing algorithm utilizes SINR metric for finding shortest routes. These routes satisfy minimum SINR constraints of links for overall energy minimization in network and automatically detour from congested areas of network. The distributed version based on local information is also explained. Simulation results display low energy consumption and low flow admission blocking probability.

4.2.6 Hierarchical Routing and Clustering

Hierarchical routing has hold importance especially in mobile ad-hoc networks but its applicability to mesh networks has been limited. One possible reason for this could the fact that most of the hierarchical routing protocol presented in literature (Pei et al. 2000; Ramanathan and Steenstrup 1998; Thai and Won-Joo 2007) assume high mobility which is rarely a case in mesh. Instead, wireless mesh show far static behavior (at least in mesh routers) and client mobility can be usually handled by typical mobility management schemes. Though efficient accommodation of clustering schemes together with channel assignment policies can explore full available capacity, designing such mechanisms with clustering is still an open issue.

4.2.7 Multi-Radio/Channel Routing

Such routing protocol mainly utilize routing metrics derived for multi-channel environment in suitable well-known routing protocols like AODV or DSR.

- *MCR* (Kyasanur and Vaidya 2006): Multichannel routing protocol (Yang et al. 2005, 2006) uses the MCR routing metric described before with Dynamic Source Routing (DSR). Periodic information is exchanged between the nodes for announcing their fixed interface and assigned channels. This way, the resultant routing paths incur less channel switching cost and achieve best possible channel diversity to avoid intra-flow interference.
- *MCRP* (So and Vaidya 2005): Proposed multi-channel routing protocol (MCRP) assumes that nodes have only single radio which can be switched between multiple channels. In MCRP, all the nodes chosen for routing a flow are required to transmit on the same channel. Hence, channel assignment occurs on per-flow basis rather than per-link. Underlying implementation mechanism of MCRP is similar to AODV.
- *MR-LQSR* (Draves et al. 2004): MR-LQSR uses DSR as underlying protocol with WCETT metric described above. Such a metric discovers routing paths with better channel quality and diversity. The channel assignment at nodes having multiple radios is assumed to pre-established using any mechanism.

4.2.8 Multicasting Protocols

Multicasting is an important operation in a network due to its wide use and applicability. First insights about multicasting in wireless mesh networks came from Roy et al. (2006). It mentions that multicast protocols for wireless multi-hop networks (e.g. On-Demand Multicast Routing Protocol (ODMRP)) use link layer broadcast and hence require changes in the unicast routing metrics. It has been shown that in case of broadcast, link quality in backward direction should not be considered because there are no ACKs involved. Also, metric product over links of a path better reveals the overall quality of the path. It then modifies existing metrics like ETX, ETT and derives METX and SPP from Dong et al. (2005) and Banerjee and Misra (2002) for increasing multicast throughput in WMNs.

The work of Flury and Wattenhofer (2007) proposes unicast, multicast and anycast routing mechanisms that use labeling based forwarding, motivated by the observation that nodes in WMNs are connected with other nodes in their closer proximity with a higher probability, and fulfill the doubling metric property.

4.2.9 Broadcast Routing Protocols

Broadcast is a required function in multi-hop wireless networks since many protocols depend on it for forwarding of the control messages. Broadcast latency minimization protocols were developed for single-channel, single-radio and single-rate ad-hoc networks by Gandhi et al. (2003). The study of Qadir et al. (2006) studies this problem for multi-channel, multi-radio, multi-rate mesh networks. It is shown that for such multi-channel mesh network, the broadcast latency problem is NP-hard. It proposes four heuristic based centralized algorithms to construct low latency broadcast forwarding trees in wireless mesh. Simulation results prove that channel assignment mechanisms designed for unicast may not work efficiently in broadcasting and hence broadcasting should also be considered while performing channel assignment. Similarly, Song et al. (2007) present a distributed broadcast tree construction algorithm which utilizes local information only. It also takes into account the link quality and interference for broadcast protocol design. A rate selection process prior to selecting actual broadcast forwarding node is described by Wang et al. (2007b). Using this, it claims to cover maximum number of possible nodes to receive broadcast in every stage at best possible rate. Dual association with APs by clients is also proposed for broadcast load minimization by Lee et al. (2005a).

In other routing approaches, Esmailpour et al. (2007) propose a mechanism in which mobile clients associated with mesh routers, route data between themselves when the back-haul mesh routers are congested. Such mechanism can be useful when mobile clients can cooperate to build a hybrid mesh. Similarly, layer-2 routing has also been proposed which performs forwarding at link layer using MAC address. On one hand, such forwarding can be faster especially in multi-hop settings but is difficult to implement and use in heterogeneous networks. The study of Gupta et al. (2006) implemented such a layer-2 forwarding for 802.11 using Wireless Distribution System (WDS). As mentioned by Gupqing Li (2005), directional antennas can be beneficial to routing as it results into higher mesh connectivity and routing paths with lesser number of hops (Saha and Johnson 2004). On the other hand, routing protocols with directional antennas should be able to coordinate transmissions in the scheduling phase and must mitigate the deafness problem.

4.3 Network Planning and Deployment

In general, network planning and deployment problem deals with optimizing number and position of mesh routers and gateway nodes while meeting certain constraints like the traffic demand and coverage. Most of the upper layer protocols design assumes known network topology but the network deployment itself involves many design challenges. We next consider gateway and mesh router related design problems.

The gateway nodes in WMNs operate as integration points between the multi-hop wireless network and the wired network. Appropriate placement of such integration points is a critical factor in achievable system capacity. The gateway placement problem was investigated by Chandra et al. (2004), who aimed to minimize the number of gateways while guaranteeing the overall required bandwidth. The problem is formulated as a network flow problem and max-flow min-cut based greedy algorithms are presented for various link models. Clustering based approach is presented by Bejerano (2004) where nodes are divided into disjoint clusters. In the next phase, a spanning tree is formed in each cluster which is rooted at the gateway node. The study of Aoun et al. (2006) presented a similar approach in which recursive searching operation greedily tries to find dominating set until the cluster radius reaches some pre-defined cluster radius. Along the same lines, He et al. (2008) proved that the gateway placement problem in general WMN graph is NP-hard. It presented ILP formulation for the problem and proposes two heuristic algorithms which try to find degree based greedy dominating tree set partitioning and weight based greedy dominating tree set partitioning for efficient gateway placement. Most of the approaches for the gateway placement consider non-varying network topologies but many of WMNs in real world actually expand incrementally. To address this, Robinson et al. (2008) modeled gateway placement problem as a facility location problem. It presents gateway-placement algorithms which take into account contention at each gateway by considering routing paths in the network. Such an approach outperforms other approaches due to actual consideration of interference and load balancing at gateways. Gateway placement scheme of Li et al. (2007) divides the network area into a grid and chooses certain cross-points as location for gateways. The study of Lakshmanan et al. (2006) motivates the need of multiple gateway association for clients for better load balancing, fairness and security concerns. In dynamic cross-layer association process presented by Athanasiou et al. (2007), clients associate to a particular mesh router not only based on channel conditions but also current AP load and routing QoS information. Problem of gateway placement is also considered jointly with routing and scheduling by Targon et al. (2009). Here authors provide mathematical formulation to study how these individual design problems affect the gateway placement.

Gateway placement problem assumes the positions of mesh routers are known but the optimization of number and position of mesh routers in WMNs also has attracted many researchers. The study of Amaldi et al. (2008) provided an ILP formulation which selects certain candidate sites for the placement of mesh routers. It takes into consideration variety of constraints such as routing, interference, channel assignment and even rate adaptation. Similarly, So and Liang (2007) presented a formulation with non-linear constraints where objective is to minimize the number of mesh routers with proper channel configuration such that the traffic demand can be satisfied. The study of Wang et al. (2007a) provides a heuristic algorithm which tries to lower the cost of installation by reducing the number of mesh routers while meeting the coverage, connectivity and demand constraints. Along the same lines, Benyamina et al. (2008a,b) consider multiple objective network planning where overall interference level is also minimized along with low cost deployment and increased throughput. The proposed solution of Benyamina et al. (2008a) also considers fault tolerance in the case of single node failure using shared protection schemes. The same set of optimization models is further extended and compared by Benyamina et al. (2009) with better load balancing of traffic across the links of the network. Some of the topological and deployment factors which can affect routing, fairness, client coverage area etc. are analytically studied by Robinson and Knightly (2007). It shows that to provide 95 % coverage, random node deployment requires as many as twice the number nodes required in a square or a hexagonal grid placement. A novel measurement driven deployment approach is presented by Camp et al. (2006) where extensive measurements are taken before the actual deployment to understand the propagation characteristics of the environment. It claims that such measurement driven approach of deployment accurately predict the required resource provisioning and achievable network capacity. Provided steps of the measurement can be a useful guideline along with site survey to eliminate possible over-provisioning and disconnections.

Most of the topology control mechanisms (like Li et al. 2003) assume altruist node behavior. The study of Santi et al. (2006) shows that in a non-cooperative ubiquitous mesh deployment (similar to wireless community networks), node may act selfishly and destroy designer's goal of optimal topology. The study of Santi et al. (2006) introduces a game-theoretic incentive-compatible framework to encourage the selfish nodes to engage in global goal of topology formation and maintenance. Addressing important issue of backbone design, Ju and Rubin (2006) present a distributed algorithm which chooses high capacity mesh nodes in backbone for relaying while Lee et al. (2007) try to build a backbone in non-cooperative environment with selfish nodes.

4.4 Capacity and Performance Modeling

In the seminal work of capacity modeling for wireless networks, Gupta and Kumar (2000) proved that for n identical wireless nodes, throughput obtainable by each node for a randomly chosen destination is $\Theta(W/\sqrt{n \log n})$ bits/sec when nodes are located randomly and each capable of transmitting at W bits/sec. When nodes are placed optimally, with optimal traffic pattern and transmission range is optimally chosen, achievable throughput can be no more than $\Theta(W/\sqrt{n})$ bits/sec.

Followed by that Li et al. (2001) proved that 802.11 MAC is capable of achieving the theoretical maximum capacity of $O(1/\sqrt{n})$ per node in a large network with n nodes randomly placed and having random traffic pattern. It is also argued that one-hop node capacity is $O(n)$ in n node ad-hoc network. As more nodes are added to network, end-to-end routing paths also grow in terms of number of hops. In such case, average routing path length will be spatial diameter of the network $O(\sqrt{n})$. This way, overall throughput at each node will be approximately $O(n/\sqrt{n}) = O(1/\sqrt{n})$.

Probably, most applicable to mesh networks is the capacity analysis presented by Gastpar and Vetterli (2002). Here, authors consider case of relays where all the nodes except source and destination relay packets with arbitrary cooperation. In such settings, when number of nodes goes to infinity the network throughput of $O(\log n)$ can be achieved. Different from traditional ad-hoc network, nodes in mesh network forward their traffic to gateways only, creating hot-spots at gateways (Jun and Sichitiu 2003). This shows that available throughput increases with increase in number of network gateways while available capacity at each node is as low as $O(1/n)$. Per-node throughput of $O(1/n)$ is also achievable in WLANs but it is empirically observed that WMNs achieve a throughput which is often lesser than WLANs. The study of Pathak and Dutta (2009) showed that WMNs achieve per-node throughput of $O(1/\delta n)$ where δ is a factor dependent on hop-radius of the network and it converges to 3 for large WMNs. In similar network settings, Zhang and Jia (2009) proved that upon using directional antennas, WMNs can achieve a capacity of $O(\frac{\log m}{\theta})$ when $m = 2$ and $O(\frac{\log m}{\theta^2 \log(1/\theta)})$ when $m > 2$, where m is the number of antennas on each node and θ is the beamwidth of antennas.

In arbitrary networks where node locations and traffic patterns can be controlled, each interface of capable of selecting appropriate transmission power, Kyasanur and Vaidya (2005b) proves that there is a loss of network capacity when the number of interfaces per node is smaller than the number of channels. While in random networks where node locations and traffic patterns are random, it is shown that one single interface is sufficient for utilizing multiple channels as long as the number of channels is scaled as $O(\log n)$ where each channel has bandwidth of W/c. The study of Bhandari and Vaidya (2007a) extends this work to multi-channel networks with channel switching constraints. It considers two kinds of channel assignments with constraints namely adjacent (c, f) channel assignment and random (c, f) assignment. In adjacent (c, f) assignment, a node is assigned and can switch between randomly chosen f continuous channels out of c available channels. In such case, per-flow capacity of $\Theta(W\sqrt{\frac{f}{cn\log n}})$ can be achieved. While in random (c, f) assignment, a node can switch between fixed random subset of f channels. Per-flow capacity in such case is $O(W\sqrt{\frac{P_{rnd}}{n\log n}})$ where $P_{rnd} = 1 - (1 - \frac{f}{c})(1 - \frac{f}{c-1})...(1 - \frac{f}{c-f+1})$. The study of Bhandari and Vaidya (2007b) shows that when $f = \Omega(\sqrt{c})$, random (c, f) assignment yields capacity of the same order as attainable via unconstrained switching. This opens up a new direction of designing routing and scheduling mechanism which can achieve this capacity bound.

4.5 Rate Control

4.5.1 TCP for Congestion Control

As with any network using shared bandwidth of links and node resources, it is necessary in mesh networks that there exists a mechanism which can adjust the data rate of sender nodes. Traditionally, this is achieved by using TCP (Transmission Control Protocol) in most of the current communication networks. The advantage of TCP is that it not only allows rate control but also provides reliability for the messages being transferred. Instead of using any explicit mechanism for detecting congestion in network, TCP relies on packet losses as an indication of congestion. The mechanism serves well in wired networks where packet losses are mostly because of packets being dropped due to congestion. In the case of wireless networks, the packet losses can be due to inherent unreliability of wireless medium. To overcome this, a large number of TCP variant for wireless networks are proposed in literature (Lochert et al. 2007).

Most of the variants of TCP for wireless were mostly developed for single last-hop wireless link. We discuss three of these single-hop variants briefly below, assuming basic familiarity with TCP (Schiller 2000). In such settings, a mobile node is connected to an access point which forwards its data to wired backbone.

- I-TCP (Indirect TCP)—I-TCP splits the TCP connection between a mobile station and remote host at the access point into two connections. This allows the isolation of wireless link because the AP itself maintains a separate TCP connection with the remote host. The advantage comes at the cost of sacrificing the end-to-end semantics on which TCP is built.
- Snoop TCP—It allows snooping of data and acknowledgments which in turn allows their retransmission on wireless link instead of an end-to-end retransmission. Though this maintains the end-to-end semantics of TCP, it does not provide an efficient isolation of wireless link as in the case of I-TCP.
- M-TCP—Mobile TCP splits the TCP connection as in I-TCP but the acknowledgments are not generated by the device where the connection is split, instead the original acknowledgments are forwarded to corresponding destination.

There are other variants such as Freeze-TCP, TCP-spoofing etc., and their further details are provided by Schiller (2000). In the last decade, a large number of TCP variants have focuses on designing efficient congestion control mechanism for multi-hop wireless networks. Rangwala et al. (2008) has provided a logical classification of these efforts which we use to in turn describe each contribution. The classification is based on three different types of enhancements that these protocol provide: (i) using better understanding of link-layer losses, (ii) by performance improvements without circumventing link-layer losses, and (iii) using improved neighborhood management.

The ideology of the first set of variants is to identify efficiently when packet losses are induced by link layer losses (node mobility or wireless medium) instead of actual congestion, and to tune TCP accordingly. Few of the protocols of this family are listed below:

- TCP-F (A Feedback Based TCP) (Chandran et al. 1998): To overcome the issue of misinterpreting link-layer loss as a congestion loss, TCP-F informs the source by Route Failure Notification when a route disruption occurs at an intermediate node. This can be used by the source to freeze its timers and terminate retransmission until the route to the destination is re-established.
- TCP-BuS (Kim et al. 2000): In this variant, on-going packets are buffered at intermediate nodes of a route disruption and re-establishment. It introduces additional messages which tunes timer values until a new route is established to avoid going

into fast retransmission phase. The protocol is shown to be especially effective in case of high node mobility in ad-hoc networks.

- TCP-ExTh-ELFN (Holland and Vaidya 1999): It develops a metric called Expected Throughput (ExTh) that can capture the performance differences as number of hops vary, and uses the metric along with Explicit Link Failure Notification (ELFN). The combination of ExTh metric and ELFN allows better tuning of TCP for further throughput improvement.
- ATCP (Ad-hoc TCP) (Liu and Singh 2001): ATCP uses information from the intermediate nodes to put TCP sender in persistent mode when no source-destination path can be found due to mobility. On the other hand, when losses occur due to link layer error, it guarantees that the sender does not enter active congestion control phase. The protocol is implemented as separate layer service between IP and TCP layers.
- Cross-layer TCP (Yu 2004): This variant builds on the idea that routing and congestion control services should better coordinate in cross-layer manner to improve TCP performance in MANETs. Early Packet Loss Notification (EPLN) notifies the sender about lost packets while Best-Effort Ack Delivery (BEAD) module enable retransmitting acknowledgments from intermediate nodes.

The difficulty in enhancing TCP by distinguishing between types of losses is that most of such methods require some extra mechanisms where intermediate nodes take part in providing necessary information. This inherently affects the end-to-end semantics of TCP protocol, and often not favored by many practical systems. To deal with this, few other variants of TCP do not try to deal with identifying causes of losses but instead uses other methods to improve the performance.

COPAS (COntention-based PAth Selection) (Cordeiro et al. 2002) is such an approach. When nodes are static and relative neighborhood remains more or less unchanged, it has been observed that due to complex interplay of MAC and TCP, there can be significant unfairness among nodes regarding how much medium access and throughput they can achieve. To address this, Cordeiro et al. (2002) developed a protocol (named COPAS) in which TCP is made aware of MAC layer contention. COPAS uses disjoint forward and reverse paths to minimize the conflicts between TCP data and ACKs. Also, the paths are constantly changed in order to minimize MAC layer contention.

Certain TCP variants recognize that congestion in multi-hop wireless networks is a neighborhood phenomenon. They derive protocols which detect and notify the congestion in neighborhood. One such protocol, namely WCP (Wireless Control Protocol) (Rangwala et al. 2008) well represents the family of neighborhood based TCP variants. WCP is a AIMD (Additive Increase Multiplicative Decrease) rate control protocol that does not require any coordination/modification of underlying MAC protocol.

The central idea of WCP is that every congested link shares its congestion information in its neighborhood. The neighborhood of a link defined as a set of links and nodes that interfere with transmission of the link and its endpoints. To detect whether a link is congested or not, every node implements and maintains Exponentially Weighted

Fig. 4.1 With typical TCP variants, the flow between Node 4 and Node 6 would starve

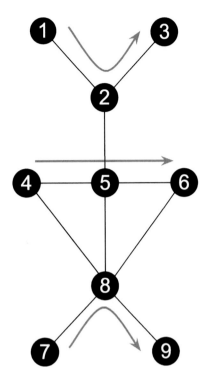

Moving Average (EWMA) for each queue for all of its outgoing links. When the average becomes greater than a preset threshold value, the link is denoted to be congested. Once a link is congested, it shares this information with all nodes in its neighborhood.

WCP utilizes AIMD as described before but the control time interval after which AIMD is executed is novel. Instead of using fixed time interval for additive increase (t_{ai}), it utilizes flow RTT values for guiding t_{ai}. Specifically, t_{ai} is set to the largest value of average shared RTT across all links from which the flow passes. This ensures that all the flows passing through a congested hot-spot increase their rates almost at the same time scale instead of following individual independent times. In multiplicative decrease, source node of a flow halves its input rate once it receives a congestion notification. The process is repeated after every t_{md} time interval. WCP sets t_{md} to shared instantaneous RTT which is similar to the shared RTT sampled at a specific time. The idea behind the choice is that instantaneous RTT well represents the current level of congestion in the network, and it more conservative compared to average shared RTT which is especially necessary to avoid aggressive reaction to congestion.

To understand the applicability and effectiveness of WCP, Rangwala et al. (2008) investigated its behavior in a topology such as that shown in Fig. 4.1. The flow between node 4 and node 6 reacts aggressively to the congestion caused by other two

Fig. 4.2 A sample topology
with one disadvantaged node
(multiple hops from the gateway)

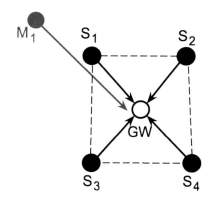

flows. In regular TCP, this results into the 4–6 flow being completely starved while
other two flows achieving very high goodput. This is avoided when WCP is used
for congestion control, since their reaction to congestion is not disproportionate as it
was in the case of TCP. In this case, all flows achieve about the same goodput; while
this is about half the goodput that flows 1–3 and 7–9 achieve under TCP, it is nearly
ten times the goodput received by 4–6.

It is worth noting that such congestion and rate control protocol can be designed
using their tight integration with network and MAC layer protocols. Such cross-layer
designs have been widely adopted even in practice, and they are discussed in further
details in Chap. 5.

4.5.2 Addressing Inherent Spatial Bias in WMNs

In a wireless mesh network, all the traffic flows are between mesh nodes and the
gateways. Due to this, nodes spatially nearer to the gateways can often achieve
better throughput as compared to nodes which are multiple hops away. This way,
single hop neighbors can in fact achieve a very high goodput, leaving multi-hop mesh
nodes starving for resources. This was further shown with an example from Mancuso
et al. (2010). In the case shown in Fig. 4.2, there are multiple one-hop neighbors
sending data to the central gateway, and there is only one traffic flows that passes
through multiple hops. Figures 4.3 and 4.4 show the throughput values that single hop
nodes and two-hop nodes can achieve using TCP and UDP respectively. The gateway
utilization is the amount of time the gateway transceiver is sending or receiving data
as opposed to being in idle state. In both cases (TCP and UCP), as the offered load
of single hop neighbors increase, the throughput of two-hop neighbor decreases.

Intuitively, the bias can be avoided if the one-hop neighbors can reduce their
input data rate. The problem with this approach is that the traffic offered by multi-
hop neighbors can dynamically change, and depending on this one-hop neighbors
can in fact under or over utilize the gateway capacity. To systematically address the

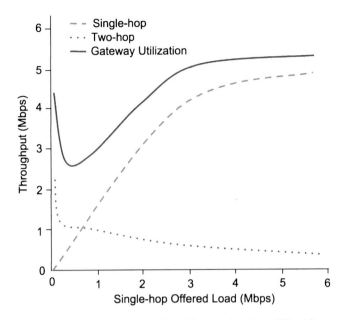

Fig. 4.3 Throughput disparity for TCP flows in topology such as that of Fig. 4.2

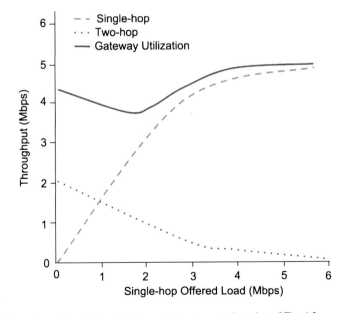

Fig. 4.4 Throughput disparity for UDP flows in topology such as that of Fig. 4.2

issue, Mancuso et al. (2010) proposed an scheme called GAP (Gateway Airtime Partitioning) which fairly partitions the gateway's capacity among mesh nodes. The GAP scheme is based on two ideas:

1. The total offered traffic (excluding relay traffic) of single-hop nodes should exceed a preset threshold. Whenever the multi-hop nodes can indeed utilize the remaining gateway airtime, single-hop nodes can further reduce their offered load below the threshold so that the resultant distribution of gateway utilization is fair
2. The utilization threshold of single-hop neighbors is elastically tuned depending on the utilization of gateway airtime by the multi-hop nodes.

Both the principles were implemented in a distributed algorithm. A total of 8 spatially distributed two hop tree branches were implemented to understand the effectiveness. Their results show that two hop neighbors can indeed achieve a comparable throughput using GAP while maintaining a good utilization of the gateway; although two-hop neighbors get a smaller share, this share does not dwindle with increasing load.

4.6 Cognitive Mesh Issues: Spectrum Sensing and Access

We have previously described the central idea behind cognitive networking—the use of spectrum by secondary users when that particular spectrum is not utilized or under-utilized by the primary user (Fig. 4.5). Naturally, the most important component of cognitive mesh network is the correct detection of activity of primary users. Further, once some spectrum is sensed to be available, there must be coordination among the secondary users as to who transmits. These problems have received significant research attention. In this section, we discuss these problems and the techniques proposed in literature.

In terms of the actual hardware required for sensing, many of the current implementations of software-defined radios such as GNU Radio-USRP (Blossom 2004; Ettus 2009), XG Radio (McHenry et al. 2007) or WARP (WARP 2010). It should be possible for the secondary users to detect primary users' communication even when primary radios use spread-spectrum technologies such as frequency hopping spread-spectrum (FHSS) or direct-sequence spread spectrum (DSSS). Other physical layer issues like detecting energy in wideband spectrum (An et al. 2011; Zhang et al. 2010; Quan et al. 2008a,b) and narrowband (Zhang and Tian 2007; Ebrahimi and Hall 2009; Zhang et al. 2008a; Yan and Gong 2010) have also attracted much research but we do not include it here considering their scope.

In general, there are multiple issues faced when detecting the spectrum activity of primary users. Two of the major issues are:

• Hidden primary user problem—secondary user can detect activities of primary users that are within its range. As it is shown in Fig. 4.6, it is possible that a primary transmitter node that is beyond the reach of secondary user transmits to

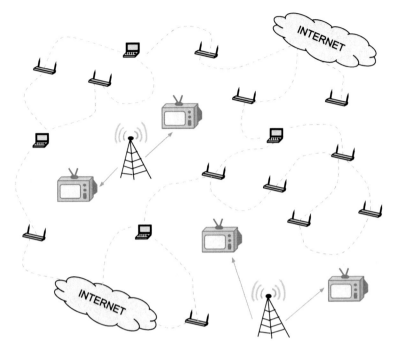

Fig. 4.5 Mesh routers operating as secondary users in holes of TV white space spectrum

Fig. 4.6 Hidden primary user
in whitespace networking

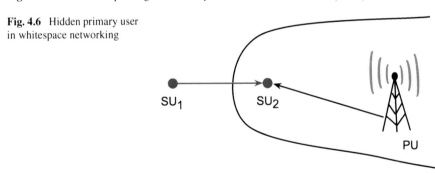

a primary receiver within the range of secondary user. This results into collision
and interference to the primary user communication.

- Uncertainty of primary user signal—it is possible in certain cases that both primary
  and secondary users are within range of each other but the activity of primary is
  not detected correctly by the secondary user. This can happen due to multi-path
  fading, shadowing or other unpredictability introduced by the wireless medium.
  In such case, secondary user might perceive that primary user is inactive, and its
  transmission may cause intolerable interference at the primary user.

Considering these issues, there are two different type of methods are used for spectrum sensing, *standalone* and *cooperative* sensing. We discuss each method in detail next.

4.6.1 Standalone Sensing

In standalone sensing, every secondary user sense the energy on channel in order to detect whether the channel is in use by the primary user. Note that noise should also be included while comparing the detected signal with a pre-determined threshold. That is, if the detected signal (d), primary user's signal is (p) and white Gaussian noise in sample is (n), then primary user is assumed to be active if $d = p + n$ and inactive if $d = n$ only.

The merit of the method lies in its simplicity. Typically, two metrics are used for evaluating the correctness of any sensing algorithm:

• Probability of false detection—this denotes the probability that a secondary user determines that a primary user is present and active when the medium is indeed free.
• Probability of correct detection—this denotes the probability that a secondary user determines that a primary user is present and active when the medium is indeed occupied by a primary user.

Typically, d is not directly used for determining the presence of primary user but instead it is compared with a pre-determined threshold. The pre-determined threshold is nothing but an SNR value used for comparison. The choice of SNR value itself affects the probabilities of false and correct detection. This was studied by Yucek and Arslan (2009), and the results are shown in Fig. 4.7. As it can be observed that higher value of SNR indeed guarantees improvement in metric probabilities.

Due to its simplicity, energy based detectors have been largely studied in research (Cabric et al. 2004, 2006; Datla et al. 2007; Digham et al. 2003; Ganesan and Li 2005; Geirhofer et al. 2006; Ghasemi and Sousa 2007; Jones et al. 2005; Lehtomki 2005; Leu et al. 2005; Liu et al. 2012; Pawelczak et al. 2006; Qihang et al. 2006; Sahai et al. 2006; Shankar et al. 2005; Tang 2005; Weidling et al. 2005; Weiss 2003; Xie et al. 2010; Yuan et al. 2007). An interested reader can refer to Yucek and Arslan (2009) for further details of standalone sensing methods.

Standalone sensing methods have many disadvantages in terms of its accuracy and efficiency. It is widely known that they are unable to solve both *hidden primary user problem* and *uncertainty in primary user signal* problems. This is because the local knowledge of sensing has only a limited use if it is not shared with other neighboring nodes. Also, these methods do not work well in detecting spread spectrum signals (Yucek and Arslan 2006). These inefficiency has given rise to other more sophisticated methods of spectrum sensing such as cooperative sensing etc. Cooperative sensing has been proven especially useful, and we discuss it next.

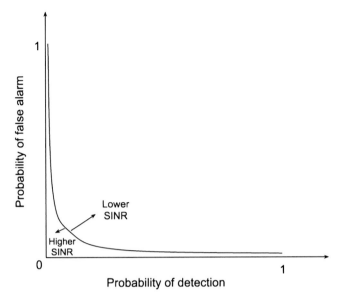

Fig. 4.7 In detecting primary users, a higher SINR threshold improves detection characteristic

4.6.2 Cooperative Sensing

The central idea of cooperative sensing is to allow nodes to cooperate by communicating their sensing observations, and improve the overall sensing performance by exploiting the spatial diversity. As compared to standalone sensing, when nodes share their sensing observations, it is likely to be possible to reach an improved consensus about the presence and activity of primary users. There are two types of cooperative sensing:

- Centralized cooperative sensing—In this type of cooperative sensing, cognitive nodes first sense the medium for activity, and then instead of making any decision, they communicate their observation to a central entity. The central entity then uses the sensing observations of difference cognitive radios to conclude if there is any activity by primary users or not. Note that there is an out-of-band control channel necessary for communication between the cognitive nodes and the central entity. The process is demonstrated in Fig. 4.8.
- Distributed cooperative sensing—In this type of cooperative sensing, there is no central entity to determine the presence of primary users, but the cognitive nodes themselves communicate their observations with each other, and distributively converge to a decision. It is also possible for each node to in fact share its local decision with others in order to remove decision error, and hopefully converge to a globally valid decision. This is shown in Fig. 4.8.

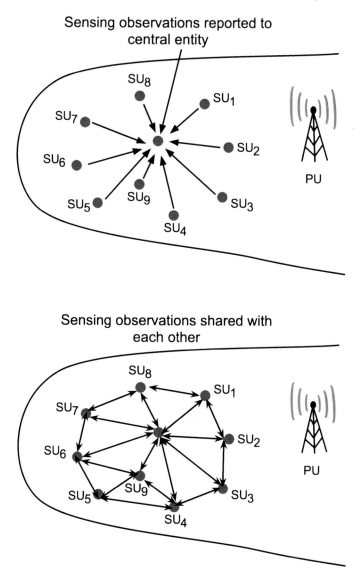

Fig. 4.8 Cooperative spectrum sensing; *top* cooperation by reporting to central entity, *bottom* distributed cooperation

Some of the examples of cooperative sensing schemes are research presented by Chen et al. (2008a), Bazerque and Giannakis (2010), Li et al. (2010), Zhang et al. (2008b) and Zheng et al. (2008). A comprehensive survey of cooperative schemes is presented by Akyildiz et al. (2011).

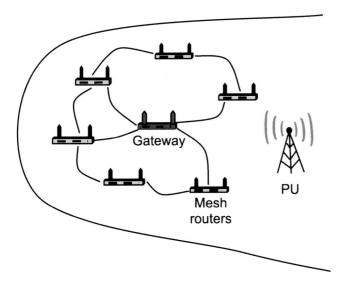

Fig. 4.9 Spectrum sensing in mesh: distinction must be made between client traffic to mesh nodes and mesh nodes to gateway traffic

4.6.3 Spectrum Sensing in Mesh Networks

Compared to an ad-hoc network, spectrum sensing is more complicated in a mesh network. This is due to two major reasons:

- Mesh networks typically consists of mesh nodes with multiple radio interfaces (at least two–one for access and other for backhaul). This requires that tunable cognitive interface should be able to control the behavior of activity of both interfaces. This is especially challenging since there is typically only one cognitive radio at each mesh node considering its cost.
- The traffic pattern of access tier and backhaul tier is quite different since access from client is sporadic while relaying in backhaul is more or less constant. This requires that sensing is adaptive to these traffic patterns in order to make them more efficient and accurate.

The issues can be better understood using Fig. 4.9. As shown, each mesh node and its clients are required to operate on a same channel. This is typically referred as a cluster by Di Felice et al. (2010, 2011) and Chowdhury et al. (2010). The other interface for the purpose of inter-cluster communication relays data from mesh nodes to a gateway. Unavailability of a channel due to primary user activity can cause temporary disconnection between mesh nodes. Whenever a channel is available, priority is given to access tier (intra-cluster links) in order to transfer as much traffic as possible from clients to mesh nodes. This is because mesh nodes are typically resource rich, and can store the data temporarily until a channel is available for inter-cluster links.

4.6.4 Medium Access

Once the spectrum sensing is complete, cognitive radio node makes the decision (standalone or cooperative) about whether or not to transmit. This attempt of medium access in turn requires that there is coordination among secondary users in terms of when and who will access the medium. Broadly, there are two levels of medium access that are necessary in cognitive mesh networks: (i) secondary users first resolve any possible contention with primary users by sensing the medium for primary user activity, this is largely achieved using spectrum sensing techniques described before, (ii) depending on the primary user activity, secondary users determine whether or not they will transmit, and in turn also makes the decision about which secondary user will access the medium so that there is no collision among the secondary user transmissions.

Broadly, MAC protocols proposed for cognitive radio networks can be divided into *time-slotted medium access* protocols, and *random access* protocols.

4.6.4.1 Time-Slotted Medium Access Protocols

As we had discussed, time slotted MAC protocols typically increase the throughput and robustness but requires synchronization. The synchronization can be achieved using a central controller or in distributed fashion. Several time-slotted MAC protocols have been proposed in literature most of which achieve synchronization using a central entity. Since such designs do not scale well with increasing number of cognitive users, many recent MAC protocols use beaconing and common control channel in order to determine a transmission schedule.

In a typical time-slotted protocol for secondary users, beaconing, neighbor discovery and exchange of other control information happens on common control channel. Typically, such a common control channel is dedicated and out-of-band to allow reliable exchange of control information. During beaconing period, nodes broadcast their sensing observation, and try to claim an inactive primary user channel for their communication. The order in which claims are made typically determine the effectiveness of protocols. After beaconing, nodes are expected to come to a consensus about the time and channel in which each cognitive radio transmits data.

802.22 (Stevenson et al. 2009; Cordeiro et al. 2005) MAC protocol is an example of such time-slotted MAC that utilizes base-station for coordinating between the cognitive radio nodes and their transmissions. Timmers et al. (2010) propose a MMAC protocol which is a distributed MAC protocol and does not require a central entity to coordinate the transmission. C-MAC (Cordeiro and Challapali 2007) uses two distinct channels (one for control and one for backup) for synchronization and transmission of information from secondary users. Similar protocol has been also presented by Chen et al. (2008b) where nodes are divided into clusters.

4.6.4.2 Random Access Protocols

Other family of protocols studied for medium access of secondary mesh users depend on random medium access mechanisms such as CSMA, ALOHA etc. Many of such protocols are surveyed by Cormio and Chowdhury (2009). The central strategy of designing a random access protocol cognitive mesh network is to manipulate the access priority of secondary and primary users. This is achievable using strategies similar to the ones implemented in IEEE 802.11e (Xiao 2005). Here, primary users have been assigned a shorter carrier sensing time by reducing the size of contention window. On the other hand, primary users are assigned a larger contention window to increase their medium access time. This provides probabilistic guarantees for each class of users and their achievable throughput. Lien et al. (2008) provides an extensive analysis of such MAC priority based mechanisms and their throughput performance. Other protocols of similar characteristics are presented by Luo and Roy (2007), Ma et al. (2005, 2007), Pawelczak et al. (2005) and Jia et al. (2008).

Recently, Tao Jing and Cheng (2012) presented an analysis that provides analysis of achievable transmission capacity of a secondary cognitive radio mesh network under three MAC strategies (CSMA, ALOHA, TDMA) when the primary network is operating along with a primary cellular network. It was concluded by Tao Jing and Cheng (2012) that cognitive mesh network achieves the maximum capacity under TDMA while the lowest capacity under ALOHA. The underlying characteristics of MAC protocols remain more or less same—higher throughput with TDMA under higher load, and higher throughput with random strategies under lower load—even in case of cognitive mesh network.

As can be seen from design of MAC layer protocols, upper layer protocols for cognitive mesh networks requires tight coordination with physical layer sensing mechanism. This requires that all upper layer protocols to be cross-layer by nature, and sensing information is expected to be shared between layers. To this end, we leave the discussion of routing and congestion control protocols of cognitive mesh networks to Chap. 5 where cross-layer design is discussed in details.

References

Abolhasan M, Hagelstein B, Wang JCP (2009) Real-world performance of current proactive multi-hop mesh protocols. In: IEEE Asia-Pacific conference on communications (APCC)

Akyildiz IF, Lee WY, Vuran MC, Mohanty S (2006) Next generation/dynamic spectrum access/cognitive radio wireless networks: a survey. Comput Netw 50(13):2127–2159. http://dx.doi.org/10.1016/j.comnet.2006.05.001

Akyildiz IF, Lo BF, Balakrishnan R (2011) Cooperative spectrum sensing in cognitive radio networks: a survey. Phys Commun 4(1):40–62. doi:10.1016/j.phycom.2010.12.003, http://www.sciencedirect.com/science/article/pii/S187449071000039X

Amaldi E, Capone A, Cesana M, Filippini I, Malucelli F (2008) Optimization models and methods for planning wireless mesh networks. Comput Netw 52(11):2159–2171

An C, Si P, Ji H (2011) Wideband spectrum sensing scheme in cognitive radio networks with multiple primary networks. In: Wireless communications and networking conference (WCNC), 2011 IEEE, pp 67–71. doi:10.1109/WCNC.2011.5779108

Aoun B, Boutaba R, Iraqi Y, Kenward G (2006) Gateway placement optimization in wireless mesh networks with QoS constraints. IEEE J Sel Areas Commun 24(11):2127–2136. doi:10.1109/JSAC.2006.881606

Athanasiou G, Korakis T, Ercetin O, Tassiulas L (2007) Dynamic cross-layer association in 802.11-based mesh networks. In: INFOCOM 2007. 26th IEEE international conference on computer communications. IEEE, pp 2090–2098. doi:10.1109/INFCOM.2007.242

Babel (2007) Babel—a loop-free distance-vector routing protocol. http://www.pps.jussieu.fr/jch/software/babel

Banerjee S, Misra A (2002) Minimum energy paths for reliable communication in multi-hop wireless networks. In: MobiHoc'02: proceedings of the 3rd ACM international symposium on mobile ad hoc networking & computing. ACM, New York, pp 146–156. http://doi.acm.org/10.1145/513800.513818

BATMAN (2007) B.A.T.M.A.N.—better approach to mobile ad-hoc networking. http://www.open-mesh.net

Bazerque J, Giannakis G (2010) Distributed spectrum sensing for cognitive radio networks by exploiting sparsity. IEEE Trans Signal Process 58(3):1847–1862. doi:10.1109/TSP.2009.2038417

Bejerano Y (2004) Efficient integration of multihop wireless and wired networks with QoS constraints. IEEE/ACM Trans Netw 12(6):1064–1078. http://dx.doi.org/10.1109/TNET.2004.838599

Benyamina D, Hafid A, Gendreau M (2008a) A multi-objective optimization model for planning robust and least interfered wireless mesh networks. In: Global telecommunications conference, 2008. IEEE GLOBECOM 2008. IEEE, pp 1–6. doi:10.1109/GLOCOM.2008.ECP.1014

Benyamina D, Hafid A, Gendreau M (2008b) Wireless mesh network planning: a multi-objective optimization approach. In: Broadband communications, networks and systems, 2008. BROAD-NETS 2008. 5th international conference on, pp 602–609. doi:10.1109/BROADNETS.2008.4769149

Benyamina D, Hafid A, Gendreau M, Hallam N (2009) Optimization models for planning wireless mesh networks: a comparative study. In: Wireless communications and networking conference, 2009. WCNC 2009. IEEE, pp 1–6. doi:10.1109/WCNC.2009.4917871

Bhandari V, Vaidya N, (2007a) Connectivity and capacity of multi-channel wireless networks with channel switching constraints. In: INFOCOM, 2007. 26th IEEE international conference on computer communications. IEEE, pp 785–793. doi:10.1109/INFCOM.2007.97

Bhandari V, Vaidya NH (2007b) Capacity of multi-channel wireless networks with random (c, f) assignment. In: MobiHoc'07: proceedings of the 8th ACM international symposium on mobile ad hoc networking and computing. ACM, New York, pp 229–238. http://doi.acm.org/10.1145/1288107.1288139

Biswas S, Morris R (2005) ExOR: opportunistic multi-hop routing for wireless networks. SIG-COMM Comput Commun Rev 35(4):133–144. http://doi.acm.org/10.1145/1090191.1080108

Blossom E (2004) Gnu radio: tools for exploring the radio frequency spectrum. Linux J 2004(122):4. http://dl.acm.org/citation.cfm?id=993247.993251

Bruno R, Conti M, Gregori E (2005) Mesh networks: commodity multihop ad hoc networks. IEEE Commun Mag 43(3):123–131. doi:10.1109/MCOM.2005.1404606

Cabric D, Mishra S, Brodersen R (2004) Implementation issues in spectrum sensing for cognitive radios. In: Signals, systems and computers, 2004. Conference record of the thirty-eighth Asilomar conference on, vol 1, pp 772–776. doi:10.1109/ACSSC.2004.1399240

Cabric D, Tkachenko A, Brodersen R (2006) Spectrum sensing measurements of pilot, energy, and collaborative detection. In: Military communications conference, 2006. MILCOM 2006. IEEE, pp 1–7. doi:10.1109/MILCOM.2006.301994

Camp J, Robinson J, Steger C, Knightly E (2006) Measurement driven deployment of a two-tier urban mesh access network. In: MobiSys'06: proceedings of the 4th international conference on mobile systems, applications and services. ACM, New York, pp 96–109. http://doi.acm.org/10.1145/1134680.1134691

Chachulski S, Jennings M, Katti S, Katabi D (2007) Trading structure for randomness in wireless opportunistic routing. In: SIGCOMM'07: proceedings of the 2007 conference on applications, technologies, architectures, and protocols for computer communications. ACM, New York, pp 169–180. http://doi.acm.org/10.1145/1282380.1282400

Chandra R, Qiu L, Jain K, Mahdian M (2004) Optimizing the placement of integration points in multi-hop wireless networks. In: ICNP'04: proceedings of the 12th IEEE international conference on network protocols. IEEE Computer Society, Washington, pp 271–282

Chandran K, Ragbunathan S, Venkatesan S, Prakash R (1998) A feedback based scheme for improving TCP performance in ad-hoc wireless networks. In: Distributed computing systems, 1998. Proceedings. 18th international conference on, pp 472–479. doi:10.1109/ICDCS.1998.679778

Chen R, Park JM, Bian K, (2008a) Robust distributed spectrum sensing in cognitive radio networks. In: INFOCOM 2008. The 27th conference on computer communications. IEEE, pp 1876–1884. doi:10.1109/INFOCOM.2008.251

Chen T, Zhang H, Matinmikko M, Katz M (2008b) Cogmesh: cognitive wireless mesh networks. In: GLOBECOM workshops, 2008 IEEE, pp 1–6. doi:10.1109/GLOCOMW.2008.ECP.37

Cheng BN, Yuksel M, Kalyanaraman S (2006) Orthogonal rendezvous routing protocol for wireless mesh networks. In: Network protocols, 2006 ICNP'06, proceedings of the 2006, 14th IEEE international conference on, pp 106–115. doi:10.1109/ICNP.2006.320204

Chowdhury K, Di Felice M, Bononi L (2010) Coral: Spectrum aware admission policy in cognitive radio mesh networks. In: Global telecommunications conference (GLOBECOM 2010), 2010 IEEE, pp 1–6. doi:10.1109/GLOCOM.2010.5683284

Clausen T, Jacquet P (2003) Optimized link state routing protocol (OLSR)

Cordeiro C, Challapali K (2007) C-mac: a cognitive mac protocol for multi-channel wireless networks. In: New frontiers in dynamic spectrum access networks, 2007. DySPAN 2007. 2nd IEEE international symposium on, pp 147–157. doi:10.1109/DYSPAN.2007.27

Cordeiro C, Challapali K, Birru D, Shankar NS (2005) IEEE 802.22: the first worldwide wireless standard based on cognitive radios. In: New frontiers in dynamic spectrum access networks, 2005. DySPAN 2005. 2005 first IEEE international symposium on, pp 328–337. doi:10.1109/DYSPAN.2005.1542649

Cordeiro C, Das S, Agrawal D (2002) COPAS: dynamic contention-balancing to enhance the performance of TCP over multi-hop wireless networks. In: Computer communications and networks, 2002. Proceedings. Eleventh international conference on, pp 382–387. doi:10.1109/ICCCN.2002.1043095

Cormio C, Chowdhury KR (2009) A survey on MAC protocols for cognitive radio networks. Ad Hoc Netw 7(7):1315–1329. doi:10.1016/j.adhoc.2009.01.002, http://www.sciencedirect.com/science/article/pii/S1570870509000043

Couto DSJD, Aguayo D, Bicket J, Morris R (2003a) A high-throughput path metric for multi-hop wireless routing. In: MobiCom'03: proceedings of the 9th annual international conference on mobile computing and networking. ACM, New York, pp 134–146. http://doi.acm.org/10.1145/938985.939000

Couto DSJD, Aguayo D, Chambers BA, Morris R (2003b) Performance of multihop wireless networks: shortest path is not enough. SIGCOMM Comput Commun Rev 33(1):83–88. http://doi.acm.org/10.1145/774763.774776

Datla D, Rajbanshi R, Wyglinski A, Minden G (2007) Parametric adaptive spectrum sensing framework for dynamic spectrum access networks. In: New frontiers in dynamic spectrum access networks, 2007. DySPAN 2007. 2nd IEEE international symposium on, pp 482–485. doi:10.1109/DYSPAN.2007.70

Di Felice M, Chowdhury K, Kassler A, Bononi L (2011) Adaptive sensing scheduling and spectrum selection in cognitive wireless mesh networks. In: Computer communications and networks (ICCCN), 2011 proceedings of 20th international conference on, pp 1–6. doi:10.1109/ICCCN.2011.6006042

Di Felice M, Chowdhury KR, Meleis W, Bononi L (2010) To sense or to transmit: a learning-based spectrum management scheme for cognitive radiomesh networks. In: Wireless mesh networks (WIMESH 2010), 2010 fifth IEEE workshop on, pp 1–6. doi:10.1109/WIMESH.2010.5507904

Digham F, Alouini MS, Simon M (2003) On the energy detection of unknown signals over fading channels. In: Communications, 2003. ICC'03. IEEE international conference on, vol 5, pp 3575–3579. doi:10.1109/ICC.2003.1204119

Dong Q, Banerjee S, Adler M, Misra A (2005) Minimum energy reliable paths using unreliable wireless links. In: MobiHoc'05: proceedings of the 6th ACM international symposium on mobile ad hoc networking and computing. ACM, New York, pp 449–459. http://doi.acm.org/10.1145/1062689.1062744

Draves R, Padhye J, Zill B (2004) Routing in multi-radio, multi-hop wireless mesh networks. In: MobiCom'04: proceedings of the 10th annual international conference on mobile computing and networking. ACM, New York, pp 114–128. http://doi.acm.org/10.1145/1023720.1023732

Durocher S, Kranakis E, Krizanc D, Narayanan L (2008) Balancing traffic load using one-turn rectilinear routing. In: Theory and applications of models of computation: lecture notes in computer science. Springer. doi:10.1007/978-3-540-79228-4_41

Ebrahimi E, Hall P (2009) A dual port wide-narrowband antenna for cognitive radio. In: Antennas and propagation, 2009. EuCAP 2009. 3rd European conference on, pp 809–812

Esa (2000) Load balancing in dence wireless multihop networks. http://userver.ftw.at/esa/java/multihop/

Esmailpour A, Jaseemuddin M, Nasser N, Bazan O (2007) Ad-hoc path: an alternative to backbone for wireless mesh networks. In: Communications, 2007 ICC'07. IEEE international conference on, pp 3752–3757. doi:10.1109/ICC.2007.618

Ettus (2009) Universal software radio peripheral. http://www.ettus.com

Fang Q, Gao J, Guibas L, de Silva V, Zhang L (2005) Glider: gradient landmark-based distributed routing for sensor networks. In: INFOCOM 2005. 24th annual joint conference of the IEEE computer and communications societies proceedings, vol 1. IEEE, pp 339–350. doi:10.1109/INFCOM.2005.1497904

Flury R, Wattenhofer R, (2007) Routing, anycast, and multicast for mesh and sensor networks. In: INFOCOM 2007. 26th IEEE international conference on computer communications. IEEE, pp 946–954. doi:10.1109/INFCOM.2007.115

Gandhi R, Parthasarathy S, Mishra A (2003) Minimizing broadcast latency and redundancy in ad hoc networks. In: MobiHoc'03: proceedings of the 4th ACM international symposium on mobile ad hoc networking and computing. ACM, New York, pp 222–232. http://doi.acm.org/10.1145/778415.778442

Ganesan G, Li Y (2005) Agility improvement through cooperative diversity in cognitive radio. In: Global telecommunications conference, 2005, vol 5. GLOBECOM'05. IEEE, pp 1–5, 2505–2509. doi:10.1109/GLOCOM.2005.1578213

Ganjali Y, Keshavarzian A (2004) Load balancing in ad hoc networks: single-path routing versus multi-path routing. In: INFOCOM 2004. Twenty-third annual joint conference of the IEEE computer and communications societies, vol 2, pp 1120–1125

Gao J, Zhang L (2009) Trade-offs between stretch factor and load-balancing ratio in routing on growth-restricted graphs. IEEE Trans Parallel Distrib Syst 20(2):171–179. doi:10.1109/TPDS.2008.75

Gastpar M, Vetterli M (2002) On the capacity of wireless networks: the relay case. In: INFOCOM 2002. Twenty-first annual joint conference of the IEEE computer and communications societies proceedings, vol 3. IEEE, pp 1577–1586. doi:10.1109/INFCOM.2002.1019409

Geirhofer S, Tong L, Sadler B (2006) A measurement-based model for dynamic spectrum access in wlan channels. In: Military communications conference, 2006. MILCOM 2006. IEEE, pp 1–7. doi:10.1109/MILCOM.2006.302405

Ghasemi A, Sousa ES (2007) Optimization of spectrum sensing for opportunistic spectrum access in cognitive radio networks. In: Consumer communications and networking conference, 2007. CCNC 2007. 4th IEEE, pp 1022–1026. doi:10.1109/CCNC.2007.206

Gupta D, LeBrun J, Mohapatra P, Chuah CN (2006) WDS-based layer 2 routing for wireless mesh networks. In: WiNTECH'06: proceedings of the 1st international workshop on wireless network testbeds, experimental evaluation and characterization. ACM, New York, pp 99–100. http://doi.acm.org/10.1145/1160987.1161010

Gupta P, Kumar P (2000) The capacity of wireless networks. IEEE Trans Inf Theory 46(2):388–404. doi:10.1109/18.825799

He B, Xie B, Agrawal DP (2008) Optimizing deployment of internet gateway in wireless mesh networks. Comput Commun 31(7):1259–1275. http://dx.doi.org/10.1016/j.comcom.2008.01.061

Holland G, Vaidya N (1999) Analysis of TCP performance over mobile ad hoc networks. In: Proceedings of the 5th annual ACM/IEEE international conference on mobile computing and networking, MobiCom'99. ACM, New York, pp 219–230. http://doi.acm.org/10.1145/313451.313540, http://doi.acm.org/10.1145/313451.313540

Hu X, Lee M, Saadawi T (2007) Progressive route calculation protocol for wireless mesh networks. In: Communications, 2007. ICC'07. IEEE international conference on, pp 4973–4978. doi:10.1109/ICC.2007.821

Hu YC, Perrig A (2004) A survey of secure wireless ad hoc routing. IEEE Secur Priv 2(3):28–39. http://doi.ieeecomputersociety.org/10.1109/MSP.2004.1

Jakllari G, Eidenbenz S, Hengartner N, Krishnamurthy S, Faloutsos M (2008) Link positions matter: a noncommutative routing metric for wireless mesh network. In: INFOCOM 2008. The 27th conference on computer communications. IEEE, pp 744–752. doi:10.1109/INFOCOM.2008.125

Jakllari G, Eidenbenz S, Hengartner N, Krishnamurthy SV, Faloutsos M (2007) Revisiting minimum cost reliable routing in wireless mesh networks. In: MobiCom'07: proceedings of the 13th annual ACM international conference on mobile computing and networking. ACM, New York, pp 302–305. http://doi.acm.org/10.1145/1287853.1287890

Jia J, Zhang Q, Shen X (2008) Hc-mac: a hardware-constrained cognitive mac for efficient spectrum management. IEEE J Sel Areas Commun 26(1):106–117. doi:10.1109/JSAC.2008.080110

Jiang W, Zhang Z, Zhong X (2007) High throughput routing in large-scale multi-radio wireless mesh networks. In: Wireless communications and networking conference, 2007 WCNC. IEEE, pp 3598–3602. doi:10.1109/WCNC.2007.659

Jing T, Huo Y, Chen X, Cheng X (2012) Achievable transmission capacity of cognitive mesh networks with different media access control. In: IEEE Infocom, 2012

Johnson D, Hu Y, Maltz D (2007) The dynamic source routing protocol (DSR) for mobile ad hoc networks for IPv4

Jones S, Merheb N, Wang IJ (2005) An experiment for sensing-based opportunistic spectrum access in csma/ca networks. In: New frontiers in dynamic spectrum access networks, 2005. DySPAN 2005. 2005 first IEEE international symposium on, pp 593–596. doi:10.1109/DYSPAN.2005.1542676

Ju HJ, Rubin I (2006) Backbone topology synthesis for multiradio mesh networks. IEEE J Sel Areas Commun 24(11):2116–2126. doi:10.1109/JSAC.2006.881611

Jun J, Sichitiu M (2003) The nominal capacity of wireless mesh networks. IEEE Wirel Commun 10(5):8–14. doi:10.1109/MWC.2003.1241089

Junhai L, Danxia Y, Liu X, Mingyu F (2009) A survey of multicast routing protocols for mobile ad-hoc networks. IEEE Commun Surv Tutor 11(1):78–91. doi:10.1109/SURV.2009.090107

Katti S, Rahul H, Hu W, Katabi D, Médard M, Crowcroft J (2006) XORs in the air: practical wireless network coding. SIGCOMM Comput Commun Rev 36(4):243–254. http://doi.acm.org/10.1145/1151659.1159942

Kim D, Toh CK, Choi Y (2000) TCP-BuS: improving TCP performance in wireless ad hoc networks. In: Communications, 2000. ICC 2000. 2000 IEEE international conference on, vol 3, pp 1707–1713. doi:10.1109/ICC.2000.853785

Korkmaz T, Zhou W (2006) On finding optimal paths in multi-radio, multi-hop mesh networks using WCETT metric. In: IWCMC'06: proceedings of the 2006 international conference on wireless communications and mobile computing. ACM, New York, pp 1375–1380. http://doi.acm.org/10.1145/1143549.1143824

Kuhn F, Wattenhofer R, Zollinger A (2003) Worst-case optimal and average-case efficient geometric ad-hoc routing. In: MobiHoc'03: proceedings of the 4th ACM international symposium on mobile ad hoc networking and computing. ACM, New York, pp 267–278. http://doi.acm.org/10.1145/778415.778447

Kwon S, Shroff N (2007) Paradox of shortest path routing for large multi-hop wireless networks. In: INFOCOM 2007, pp 1001–1009. doi:10.1109/INFCOM.2007.121

Kwon S, Shroff NB (2006) Energy-efficient interference-based routing for multi-hop wireless networks. In: INFOCOM 2006. 25th IEEE international conference on computer communications Proceedings, pp 1–12. doi:10.1109/INFOCOM.2006.199

Kyasanur P, Vaidya N, (2005a) Routing and interface assignment in multi-channel multi-interface wireless networks. In: Wireless communications and networking conference, 2005, vol 4. IEEE, pp 2051–2056. doi:10.1109/WCNC.2005.1424834

Kyasanur P, Vaidya NH (2005b) Capacity of multi-channel wireless networks: impact of number of channels and interfaces. In: MobiCom'05: proceedings of the 11th annual international conference on mobile computing and networking. ACM, New York, pp 43–57. http://doi.acm.org/10.1145/1080829.1080835

Kyasanur P, Vaidya NH (2006) Routing and link-layer protocols for multi-channel multi-interface ad hoc wireless networks. SIGMOBILE Mob Comput Commun Rev 10(1):31–43. http://doi.acm.org/10.1145/1119759.1119762

Lakshmanan S, Sundaresan K, Sivakumar R (2006) On multi-gateway association in wireless mesh networks. In: Wireless mesh networks, 2006, WiMesh 2006, 2nd IEEE workshop on, pp 64–73. doi:10.1109/WIMESH.2006.288603

Lassila P (2006) Spatial node distribution of the random waypoint mobility model with applications. IEEE Trans Mob Comput 5(6):680–694. http://dx.doi.org/10.1109/TMC.2006.86, member-Hyytia, Esa and Member-Virtamo, Jorma

Lee D, Chandrasekaran G, Sinha P (2005a) Optimizing broadcast load in mesh networks using dual association. In: IEEE workshop on wireless mesh networks, WiMesh

Lee S, Bhattacharjee B, Banerjee S (2005b) Efficient geographic routing in multihop wireless networks. In: MobiHoc'05: proceedings of the 6th ACM international symposium on mobile ad hoc networking and computing. ACM, New York, pp 230–241. http://doi.acm.org/10.1145/1062689.1062720

Lee S, Levin D, Gopalakrishnan V, Bhattacharjee B (2007) Backbone construction in selfish wireless networks. SIGMETRICS Perform Eval Rev 35(1):121–132. http://doi.acm.org/10.1145/1269899.1254896

Lee SJ, Gerla M, (2000) AODV-BR: backup routing in ad hoc networks. In: Wireless communications and networking conference, 2000. WCNC, vol 3. IEEE, pp 1311–1316. doi:10.1109/WCNC.2000.904822

Lehtomki J (2005) Analysis of energy based signal detection. Oulu University Library, OAI repository test [http://herkulesoulufi/OAI/OAI-scriptxsql] (Finland), http://herkules.oulu.fi/isbn9514279255/

Leu A, Steadman K, McHenry M, Bates J (2005) Ultra sensitive tv detector measurements. In: New frontiers in dynamic spectrum access networks, 2005. DySPAN 2005. 2005 first IEEE international symposium on, pp 30–36. doi:10.1109/DYSPAN.2005.1542614

Li F, Wang Y (2008a) Circular sailing routing for wireless networks. In: INFOCOM 2008. The 27th conference on computer communications. IEEE, pp 1346–1354: doi:10.1109/INFOCOM.2008.192

Li F, Wang Y (2008b) Stretch factor of curveball routing in wireless network: cost of load balancing. In: Communications, 2008, ICC'08. IEEE international conference on, pp 2650–2654. doi:10.1109/ICC.2008.501

Li F, Wang Y, Li XY (2007) Gateway placement for throughput optimization in wireless mesh networks. In: Communications, 2007. ICC'07. IEEE international conference on, pp 4955–4960. doi:10.1109/ICC.2007.818

Li G, Steven W, Yang L (2005) Opportunities and challenges in mesh networks using directional anntenas. In: Proceedings of IEEE workshop on wireless mesh networks, WiMesh

Li J, Blake C, Couto DSD, Lee HI, Morris R (2001) Capacity of ad hoc wireless networks. In: MobiCom'01: proceedings of the 7th annual international conference on mobile computing and networking. ACM, New York, pp 61–69. http://doi.acm.org/10.1145/381677.381684

Li N, Hou J, Sha L (2003) Design and analysis of an MST-based topology control algorithm. In: INFOCOM 2003. Twenty-second annual joint conference of the IEEE computer and communications societies, vol 3. IEEE, pp 1702–1712

Li Z, Yu F, Huang M (2010) A distributed consensus-based cooperative spectrum-sensing scheme in cognitive radios. IEEE Trans Veh Technol 59(1):383–393. doi:10.1109/TVT.2009.2031181

Lien SY, Tseng CC, Chen KC (2008) Carrier sensing based multiple access protocols for cognitive radio networks. In: Communications, 2008. ICC'08. IEEE international conference on, pp 3208–3214. doi:10.1109/ICC.2008.604

Liu S, Lazos L, Krunz M (2012) Cluster-based control channel allocation in opportunistic cognitive radio networks. IEEE Trans Mob Comput PP(99):1. doi:10.1109/TMC.2012.33

Liu J, Singh S (2001) ATCP: TCP for mobile ad hoc networks. IEEE J Sel Areas Commun 19(7):1300–1315. doi:10.1109/49.932698

Lochert C, Scheuermann B, Mauve M (2007) A survey on congestion control for mobile ad hoc networks: research articles. Wirel Commun Mob Comput 7:655–676. doi:10.1002/wcm.v7:5, http://dl.acm.org/citation.cfm?id=1255143.1255152

Luo L, Roy S (2007) Analysis of search schemes in cognitive radio. In: Networking technologies for software define radio networks, 2007, 2nd IEEE workshop on, pp 17–24. doi:10.1109/SDRN.2007.4348969

Ma L, Han X, Shen CC (2005) Dynamic open spectrum sharing mac protocol for wireless ad hoc networks. In: New frontiers in dynamic spectrum access networks, 2005. DySPAN 2005. 2005 first IEEE international symposium on, pp 203–213. doi:10.1109/DYSPAN.2005.1542636

Ma L, Shen CC, Ryu B (2007) Single-radio adaptive channel algorithm for spectrum agile wireless ad hoc networks. In: New frontiers in dynamic spectrum access networks, 2007. DySPAN 2007. 2nd IEEE international symposium on, pp 547–558. doi:10.1109/DYSPAN.2007.78

Mancuso V, Gurewitz O, Khattab A, Knightly E (2010) Elastic rate limiting for spatially biased wireless mesh networks. In: INFOCOM, 2010 proceedings IEEE, pp 1–9. doi:10.1109/INFCOM.2010.5461991

McHenry M, Livsics E, Nguyen T, Majumdar N (2007) Xg dynamic spectrum sharing field test results. In: New frontiers in dynamic spectrum access networks, 2007. DySPAN 2007. 2nd IEEE international symposium on, pp 676–684. doi:10.1109/DYSPAN.2007.90

Mei A, Stefa J (2008) Routing in outer space. In: INFOCOM 2008. The 27th conference on computer communications. IEEE, pp 2234–2242. doi:10.1109/INFOCOM.2008.291

Nandiraju NS, Nandiraju DS, Agrawal DP (2006) Multipath routing in wireless mesh networks. In: Mobile adhoc and sensor systems (MASS), 2006 IEEE international conference on, pp 741–746. doi:10.1109/MOBHOC.2006.278644

Nandiraju N, Nandiraju D, Santhanam L, He B, Wang J, Agrawal D (2007) Wireless mesh networks: current challenges and future directions of web-in-the-sky. IEEE Wirel Commun 14(4):79–89. doi:10.1109/MWC.2007.4300987

Pathak P, Dutta R (2009) Impact of power control on capacity of large scale wireless mesh networks. In: IEEE ANTS 2009. doi:10.1109/ANTS.2009.5409865

Pawelczak P, Janssen G, Prasad R (2006) Wlc10-4: performance measures of dynamic spectrum access networks. In: Global telecommunications conference, 2006. GLOBECOM'06. IEEE, pp 1–6. doi:10.1109/GLOCOM.2006.671

Pawelczak P, Venkatesha Prasad R, Xia L, Niemegeers I (2005) Cognitive radio emergency networks–requirements and design. In: New frontiers in dynamic spectrum access networks, 2005. DySPAN 2005. 2005 first IEEE international symposium on, pp 601–606. doi:10.1109/DYSPAN.2005.1542678

Pei G, Gerla M, Hong X (2000) LANMAR: landmark routing for large scale wireless ad hoc networks with group mobility. In: MobiHoc'00: proceedings of the 1st ACM international symposium on mobile ad hoc networking and computing. IEEE Press, Piscataway, pp 11–18

Perkins C, Belding-Royer E, Das S (2003) Ad hoc on-demand distance vector (AODV) routing

Pham P, Perreau S (2003) Performance analysis of reactive shortest path and multipath routing mechanism with load balance. In: INFOCOM 2003. Twenty-second annual joint conference of the IEEE computer and communications societies, vol 1. IEEE, pp 251–259

Popa L, Rostamizadeh A, Karp R, Papadimitriou C, Stoica I (2007) Balancing traffic load in wireless networks with curveball routing. In: MobiHoc'07: proceedings of the 8th ACM international symposium on mobile ad hoc networking and computing. ACM, New York, pp 170–179. http://doi.acm.org/10.1145/1288107.1288131

Qadir J, Misra A, Chou CT (2006) Minimum latency broadcasting in multi-radio multi-channel multi-rate wireless meshes. In: Sensor and ad hoc communications and networks, 2006 SECON'06, 3rd annual IEEE communications society on, vol 1, pp 80–89. doi:10.1109/SAHCN.2006.288412

Qihang P, Kun Z, Jun W, Shaoqian L (2006) A distributed spectrum sensing scheme based on credibility and evidence theory in cognitive radio context. In: Personal, indoor and mobile radio communications, 2006 IEEE 17th international symposium on, pp 1–5. doi:10.1109/PIMRC.2006.254365

Quan Z, Cui S, Sayed A, Poor H (2008a) Wideband spectrum sensing in cognitive radio networks. In: Communications, 2008. ICC'08. IEEE international conference on, pp 901–906. doi:10.1109/ICC.2008.177

Quan Z, Cui S, Sayed A, VincentPoor H (2008b) Spatial-spectral joint detection for wideband spectrum sensing in cognitive radio networks. In: Acoustics, speech and signal processing, 2008. ICASSP 2008. IEEE international conference on, pp 2793–2796. doi:10.1109/ICASSP.2008.4518229

Ramachandran K, Buddhikot M, Chandranmenon G, Miller S, Belding-Royer E, Almeroth K (2005) On the design and implementation of infrastructure mesh networks. In: Wireless mesh networks, 2005 WiMesh 2005, 1st IEEE workshop on

Ramanathan R, Steenstrup M (1998) Hierarchically-organized, multihop mobile wireless networks for quality-of-service support. Mob Netw Appl 3(1):101–119. http://dx.doi.org/10.1023/A:1019148009641

Rangwala S, Jindal A, Jang KY, Psounis K, Govindan R (2008) Understanding congestion control in multi-hop wireless mesh networks. In: Proceedings of the 14th ACM international conference on mobile computing and networking, MobiCom'08. ACM, New York, pp 291–302. http://doi.acm.org/10.1145/1409944.1409978, http://doi.acm.org/10.1145/1409944.1409978

Robinson J, Knightly E (2007) A performance study of deployment factors in wireless mesh networks. In: INFOCOM 2007. 26th IEEE international conference on computer communications. IEEE, pp 2054–2062. doi:10.1109/INFCOM.2007.238

Robinson J, Uysal M, Swaminathan R, Knightly E (2008) Adding capacity points to a wireless mesh network using local search. In: INFOCOM 2008. The 27th conference on computer communications. IEEE, pp 1247–1255. doi:10.1109/INFOCOM.2008.181

Roy S, Koutsonikolas D, Das S, Hu YC (2006) High-throughput multicast routing metrics in wireless mesh networks. In: ICDCS'06: proceedings of the 26th IEEE international conference on distributed computing systems. IEEE Computer Society, Washington, p 48. http://dx.doi.org/10.1109/ICDCS.2006.46

Rozner E, Seshadri J, Mehta Y, Qiu L (2006) Simple opportunistic routing protocol for wireless mesh networks. In: Wireless mesh networks, 2006 WiMesh, 2006 2nd IEEE workshop on, pp 48–54. doi:10.1109/WIMESH.2006.288602

Saha A, Johnson D (2004) Routing improvement using directional antennas in mobile ad hoc networks. In: Global telecommunications conference, 2004. GLOBECOM'04, vol 5. IEEE, pp 2902–2908. doi:10.1109/GLOCOM.2004.1378885

Sahai A, Tandra R, Mishra SM, Hoven N (2006) Fundamental design tradeoffs in cognitive radio systems. In: Proceedings of the first international workshop on technology and policy for accessing spectrum, TAPAS'06. ACM, New York. doi:10.1145/1234388.1234390, http://doi.acm.org/10.1145/1234388.1234390

Santi P, Eidenbenz S, Resta G (2006) A framework for incentive compatible topology control in non-cooperative wireless multi-hop networks. In: DIWANS'06: proceedings of the 2006 workshop on dependability issues in wireless ad hoc networks and sensor networks. ACM, New York, pp 9–18. http://doi.acm.org/10.1145/1160972.1160976

Schiller J (2000) Mobile communications. Addison-Wesley Longman Publishing Co., Inc., Boston

Sengupta S, Rayanchu S, Banerjee S (2007) An analysis of wireless network coding for unicast sessions: the case for coding-aware routing. In: INFOCOM 2007. 26th IEEE international conference on computer communications. IEEE, pp 1028–1036. doi:10.1109/INFCOM.2007.124

Shankar N, Cordeiro C, Challapali K (2005) Spectrum agile radios: utilization and sensing architectures. In: New frontiers in dynamic spectrum access networks, 2005. DySPAN 2005. 2005 first IEEE international symposium on, pp 160–169. doi:10.1109/DYSPAN.2005.1542631

So A, Liang B (2007) Minimum cost configuration of relay and channel infrastructure in heterogeneous wireless mesh networks. In: Networking, pp 275–286

So J, Vaidya NH (2005) Routing and channel assignment in multi-channel multi-hop wireless networks with single network interface. In: The second international conference on quality of service in heterogeneous wired/wireless networks (QShine)

Song M, Wang J, Hao Q (2007) Broadcasting protocols for multi-radio multi-channel and multi-rate mesh networks. In: Communications, 2007. ICC'07. IEEE international conference on, pp 3604–3609. doi:10.1109/ICC.2007.594

Stevenson C, Chouinard G, Lei Z, Hu W, Shellhammer S, Caldwell W (2009) IEEE 802.22: the first cognitive radio wireless regional area network standard. IEEE Commun Mag 47(1):130–138. doi:10.1109/MCOM.2009.4752688

Subramanian A, Buddhikot M, Miller S (2006) Interference aware routing in multi-radio wireless mesh networks. In: Wireless mesh networks, 2006 WiMesh 2006, 2nd IEEE workshop on, pp 55–63. doi:10.1109/WIMESH.2006.288620

Subramanian S, Shakkottai S, Gupta P (2007) On optimal geographic routing in wireless networks with holes and non-uniform traffic. In: INFOCOM 2007. 26th IEEE international conference on computer communications. IEEE, pp 1019–1027. doi:10.1109/INFCOM.2007.123

Subramanian S, Shakkottai S, Gupta P (2008) Optimal geographic routing for wireless networks with near-arbitrary holes and traffic. In: INFOCOM 2008. The 27th conference on computer communications. IEEE, pp 1328–1336. doi:10.1109/INFOCOM.2008.190

Tang H (2005) Some physical layer issues of wide-band cognitive radio systems. In: New frontiers in dynamic spectrum access networks, 2005. DySPAN 2005. 2005 first IEEE international symposium on, pp 151–159. doi:10.1109/DYSPAN.2005.1542630

Tang S, Suzuki R, Obana S (2007) An opportunistic progressive routing (OPR) protocol maximizing channel efficiency. In: Global telecommunications conference, 2007. GLOBECOM'07. IEEE, pp 1285–1290

Targon V, Sans B, Capone A (2009) The joint gateway placement and spatial reuse problem in wireless mesh networks. Comput Netw 6:31–33

Thai PN, Won-Joo H (2007) Hierarchical routing in wireless mesh network. In: Advanced communication technology, the 9th international conference on, vol 2, pp 1275–1280. doi:10.1109/ICACT.2007.358590

Timmers M, Pollin S, Dejonghe A, Van der Perre L, Catthoor F (2010) A distributed multichannel mac protocol for multihop cognitive radio networks. IEEE Trans Veh Technol 59(1):446–459. doi:10.1109/TVT.2009.2029552

Tsai J, Moors T (2006) A review of multipath routing protocols: from wireless ad hoc to mesh networks. In: ACoRN early career researcher workshop on wireless multihop networking

Wang J, Xie B, Cai K, Agrawal D (2007a) Efficient mesh router placement in wireless mesh networks. In: Mobile adhoc and sensor systems. MASS 2007. IEEE internatonal conference on, pp 1–9. doi:10.1109/MOBHOC.2007.4428616

Wang T, Du X, Cheng W, Yang Z, Liu W (2007b) A fast broadcast tree construction in multi-rate wireless mesh networks. In: Communications, 2007. ICC'07. IEEE international conference on, pp 1722–1727. doi:10.1109/ICC.2007.288

WARP (2010) Rice university WARP project. http://warp.rice.edu

Weidling F, Datla D, Petty V, Krishnan P, Minden G (2005) A framework for r.f. spectrum measurements and analysis. In: New frontiers in dynamic spectrum access networks, 2005. DySPAN 2005. 2005 first IEEE international symposium on, pp 573–576. doi:10.1109/DYSPAN.2005.1542672

Weiss T (2003) A diversity approach for the detection of idle spectral resources in spectrum pooling systems. In: Proceedings of 48th international scientific colloquium, Sept 2003. http://ci.nii.ac.jp/naid/10019709795/en/

Xiao Y (2005) Performance analysis of priority schemes for IEEE 802.11 and IEEE 802.11e wireless lans. IEEE Wirel Commun Trans 4(4):1506–1515. doi:10.1109/TWC.2005.850328

Xie S, Liu Y, Zhang Y, Yu R (2010) A parallel cooperative spectrum sensing in cognitive radio networks. IEEE Trans Veh Technol 59(8):4079–4092. doi:10.1109/TVT.2010.2056943

Yan Y, Gong Y (2010) Energy detection of narrowband signals in cognitive radio systems. In: Wireless communications and signal processing (WCSP), 2010 international conference on, pp 1–5. doi:10.1109/WCSP.2010.5633150

Yang Y, Wang J, Kravets R (2005) Designing routing metrics for mesh networks. In: IEEE workshop on wireless mesh networks, WiMesh

Yang Y, Wang J, Kravets R (2005) Interference-aware load balancing for multihop wireless networks. Technical report, University of Illinois, Urbana-Champaign. http://www.cs.uiuc.edu/research/techreports.php?report=UIUCDCS-R-2005-2526

Yang Y, Wang J, Kravets R (2006) Load-balanced routing for mesh networks. SIGMOBILE Mob Comput Commun Rev 10(4):3–5. doi:10.1145/1215976.1215979, http://doi.acm.org/10.1145/1215976.1215979

Yu X (2004) Improving TCP performance over mobile ad hoc networks by exploiting cross-layer information awareness. In: Proceedings of the 10th annual international conference on mobile computing and networking, MobiCom'04. ACM, New York, pp 231–244. http://doi.acm.org/10.1145/1023720.1023743, http://doi.acm.org/10.1145/1023720.1023743

Yuan Y, Bahl P, Chandra R, Chou P, Ferrell J, Moscibroda T, Narlanka S, Wu Y (2007) Knows: cognitive radio networks over white spaces. In: New frontiers in dynamic spectrum access networks, 2007. DySPAN 2007. 2nd IEEE international symposium on, pp 416–427. doi:10.1109/DYSPAN.2007.61

Yuan Y, Wong SHY, Lu S, Arbuagh W (2005) ROMER: resilient opportunistic mesh routing for wireless mesh networks. In: Wireless mesh networks, 2005 WiMesh 2005, 1st IEEE workshop on

Yucek T, Arslan H (2006) Spectrum characterization for opportunistic cognitive radio systems. In: Military communications conference, 2006, MILCOM 2006. IEEE, pp 1–6. doi:10.1109/MILCOM.2006.302124

Yucek T, Arslan H (2009) A survey of spectrum sensing algorithms for cognitive radio applications. IEEE Commun Surv Tutor 11(1):116–130. doi:10.1109/SURV.2009.090109

Zhang D, Tian Z (2007) Spatial capacity of cognitive radio networks: narrowband versus ultra-wideband systems. In: Wireless communications and networking conference, 2007, WCNC 2007. IEEE, pp 6–10. doi:10.1109/WCNC.2007.7

Zhang J, Jia X (2009) Capacity analysis of wireless mesh networks with omni or directional antennas. In: INFOCOM 2009. IEEE, pp 2881–2885. doi:10.1109/INFCOM.2009.5062251

Zhang W, Mallik R, Ben Letaief K (2008b) Cooperative spectrum sensing optimization in cognitive radio networks. In: Communications, 2008. ICC'08. IEEE International Conference on, pp 3411–3415. doi:10.1109/ICC.2008.641

Zhang Z, Li H, Yang D, Pei C (2010) Space-time bayesian compressed spectrum sensing for wideband cognitive radio networks. In: New frontiers in dynamic spectrum, 2010 IEEE symposium on, pp 1–11. doi:10.1109/DYSPAN.2010.5457841

Zhang D, Tian Z, Wei G (2008) Spatial capacity of narrowband versus ultra-wideband cognitive radio systems. IEEE Wirel Commun Trans 7(11):4670–4680. doi:10.1109/T-WC.2008.070746

Zheng X, Wang J, Cui L, Chen J, Wu Q (2008) A novel cooperative spectrum sensing algorithm in cognitive radio systems. In: Wireless communications, networking and mobile computing, 2008. WiCOM'08. 4th international conference on, pp 1–4. doi:10.1109/WiCom.2008.302

Chapter 5
Joint Design

5.1 Joint Design

As we discussed earlier, it is readily apparent that various individual design problems are themselves highly interdependent; this is well understood among researchers. Over the course of several years of research, it has become obvious that dealing with these interdependent problems jointly is preferable (indeed, almost unavoidable) in optimizing performance. For example, even a highly effective link scheduling algorithm may not yield very good performance, because some links exist in the network that are heavily used and also centrally located so that they interfere with many other links. It is attractive to reach into the solution of the routing problem, and adjust routes so that such links are less utilized, and therefore need to be less frequently scheduled. This gives rise to an area of designing algorithms that *jointly* provide both link schedules and routes.

Previous surveys of literature largely pre-date this recent body of literature. Some surveys such as those by Foukalas et al. (2008) and Shariat et al. (2009) cover cross-layer design proposals but they focus on single-hop infrastructure networks only. We have ourselves surveyed this area previously (Pathak and Dutta 2010). In this chapter, we discuss the research problems that have emerged in this joint design area, and have been addressed in literature, with a brief survey of the literature. For a fuller survey, we refer the reader to Pathak and Dutta (2010).

5.2 Power Control and Scheduling

As we have seen before, scheduling algorithms must take into consideration the interference relationships between the links which in turn is decided by the power assignments at nodes. The nodes transmitting at high power level creates higher interference links which reduces the overall spatial reuse when scheduled. One of the first solutions to problem of jointly scheduling and assigning power control, with

P. H. Pathak and R. Dutta, *Designing for Network and Service Continuity in Wireless Mesh Networks*, Signals and Communication Technology, DOI: 10.1007/978-1-4614-4627-9_5, © Springer Science+Business Media New York 2013

the objective of maximizing throughput and minimizing the power consumption, was provided by ElBatt and Ephremides (2004). They provided a two-phase algorithm that is centralized and needs to be executed before every slot. In the first phase, the algorithm determines the maximum set of nodes that can transmit in a given slot with the constraint that they should be spatially separated by at least some distance to avoid mutual interference. In the second phase, such feasible sets of transmitting nodes are assigned power levels to meet their SINR constraints.

Similarly, Chen and Lee (2006) propose a two phase distributed algorithm for power control and link scheduling in wireless networks with the objective of through-put enhancement by lowering interference. In the first phase, all nodes having data to send first probe the channel with some initial predetermined power by sending probe packets and measures the interference before (thermal noise) and after (interference from others) the probe. Using the value of increased interference, the node calculates the SNR of the link. If its SNR is above a certain threshold then it is scheduled in the next time slot. All the links whose SNR is too low are marked undetermined and left for future scheduling. The feasible set of links run a power optimization algorithm to optimize their power for transmission. Undetermined links still check if they can be a part of the schedule after feasible links use optimal power levels, and join the schedule if they can.

A scheduling protocol should try to schedule as many links as possible in every slot of schedule to reduce the overall schedule length. The study of Moscibroda and Wattenhofer (2006) defines the notion of *scheduling complexity*, the amount of time required to schedule a given set of requests, and uses it to analyze the capac-ity of wireless networks. It argues that even in case of large networks, there is no fundamental scalability problem in scheduling the transmission requests. Scheduling protocols that use uniform or linear power assignments perform much worse in terms of the scheduling complexity. Instead, Moscibroda and Wattenhofer (2006) propose a non-linear power assignment for scheduling the links, where power assigned to a link does not directly depend on its length. Such disproportionate power assignment favors shorter links over longer links, and transmitters on the shorter links transmit at a higher power than is actually needed to reach the intended receiver. In contrast, transmitting nodes of longer links still transmit at a higher required power. Based on this non-linear power assignment, a theoretical scheduling algorithm for SINR model is presented that schedules a connected set of links.

The traditional SINR-based physical model does not capture the effects of reflection, shadowing, scattering and diffraction on radio propagation. Accordingly, Moscibroda et al. (2006) proposes a generalized physical interference model in which received signal power at the receiver can deviate from theoretically received power by the factor of f. This way, if u transmits the data to v using a transmission power P_u, α being the path-loss exponent and d_{uv} the distance between nodes u and v, then the received signal power ($P_v(u)$) at node v can range between following boundaries.

$$\frac{P_u}{f \cdot d_{uv}{}^\alpha} \le P_v(u) \le \frac{f \cdot P_u}{d_{uv}{}^\alpha} \tag{5.1}$$

The received power $P_v(u)$ should then be considered with respect to interference from other nodes at the receiver using the standard physical interference model. This extends the scheduling algorithm of Moscibroda and Wattenhofer (2006) to schedule arbitrary set of communication requests. Moscibroda et al. (2006) show that when transmission power levels are carefully chosen, the scheduling complexity of arbitrary topologies can be $O(I_{in} \cdot \log^2 n)$ with n nodes where I_{in} is a static parameter called in-interference. I_{in} is usually realized by topology control algorithms and hence topology control algorithm yielding low I_{in} achieves faster scheduling. One interesting result outlined by Moscibroda et al. (2006) is that topologies having unidirectional links yields lower I_{in} and therefore faster schedules compared to topologies having symmetric links. A combined algorithm of power assignment, topology control and scheduling is presented with a generalized physical signal propagation model.

Continuing with distance based estimation of interference, Moscibroda et al. (2007) propose a notion of *disturbance* of a link. Disturbance of a link is the larger of the maximal number of senders (or receivers) in close proximity of the sender (receiver) of the link. They propose the Low-Disturbance Scheduling protocol (LDS) which can achieve faster schedules of length within polylogarithmic factor of the network's disturbance even in worst-case low-disturbance networks. Recently, Goussevskaia et al. (2007) proved the TDMA based link scheduling problem to be NP-complete when the geometric SINR model of interference is used. In the geometric SINR model, the traditional SINR model is modified for the belief that the gain between two nodes is determined by the distance between them.

The rate at which a node injects data into the network is also an important tunable variable that can be considered together with power control mechanisms. The study by Kulkarni et al. (2004) formulates joint scheduling, power control and rate control problem as a mixed integer linear programming problem. It tries to achieve link scheduling and power assignment while meeting the data rate and peak power level constraints, such that the resulting throughput is maximized. It provides a greedy heuristic for solving the optimization problem in large networks. Capone and Carello (2006) present a joint problem where TDMA scheduling, dynamic slot-by-slot power control and transmission rate control with regards to SINR are considered. The intended transmission rate is expressed in terms of packets transmitted per slot, and SINR threshold is used to relate the rate with their corresponding SINR. Two separate formulations (a linear number of variables based model, and a column generation based model) are provided for minimizing the number of used time slots in the TDMA schedule derived, which tries to meet required SINR and traffic rate constraints.

5.3 Routing and Scheduling

Once the traffic demands are routed on specific routing paths, the scheduling algorithm tries to achieve a conflict-free schedule for links on these routing paths. If a certain link can not be scheduled with any other link in the network, the traffic on such a link could be re-routed on other routing paths. Hence, several approaches try to iteratively decide on routing paths and scheduling links to achieve a better overall throughput. Tassiulas and Ephremides (1992) explored the problem of joint routing and scheduling for packet radio networks. Because of many simplified assumptions like 1-hop interference model, the solution holds limited practical importance, but it provided an early baseline theoretical approach towards the problem. The study of Wu (2006) proposes two centralized algorithms for joint routing and scheduling which use TDMA based contention free scheduling, and utilize paths with better quality links, to fulfill the bandwidth requirement. The study uses the k-hop interference model, where any node within k hops of the receiver should not be transmitting simultaneously. It proposes an approach to estimating the value of k using the $SINR_{threshold}$ and path-loss exponent α used in SINR physical interference model as follows.

$$(\sqrt[\alpha]{SINR_{threshold}} + 1) > k \geq (\sqrt[\alpha]{SINR_{threshold}}) \qquad (5.2)$$

A heuristic approach to the LP problem formulation of integrated routing and MAC link scheduling chooses routing paths based on locally available information about the MAC bandwidth, and tries to avoid the congested areas. Interference relations between links are captured using a conflict graph derived for the above mentioned k-hop interference model.

Most current work assumes traffic information is available *a priori*, and various scheduling and routing algorithms are designed on this basis. Such assumption can be unrealistic in real-time network deployments. This motivates Wang et al. (2007) to propose a joint traffic-oblivious routing and scheduling (TORS) algorithm which can accept any or even no traffic estimation and can still provide efficient routing paths and schedules. It provides a LP formulation with no specific assumption of interference model, and utilizes the conflict graph to resolve the scheduling conflicts. The study of Bhatia and Li (2007) addresses the routing and scheduling problem for MIMO links as a cross-layer optimization problem. It also provides a LP formulation for throughput optimization with fairness constraint for physical layer resource allocation.

Lim et al. (2006) present a novel coordinate-based mechanism in which RSSI measurements between a node n and its neighbors are represented as a $p \times p$ square matrix, and each column of such a matrix can be considered as coordinates of the respective nodes in p-dimensional space. Such a virtual coordinate system can be used to find the Euclidean distance between nodes. If such a distance is large, it can be estimated that the transmissions of such nodes will not interfere with each other. This way, nodes which are not in transmission range of each other can also determine least inter-flow interference paths for routing. Once such paths are determined, the scheduling scheme allows the gateway node to transmit for longer than other mesh nodes, reasonably assuming that it has higher traffic demand. This provides better chances of scheduling multiple transmissions simultaneously, exploiting their temporal-spatial diversity. The study of Zhang et al. (2005) formulates a joint routing and scheduling problem for multi-radio multi-channel mesh, and finds the *concurrent transmission pattern*, which signifies transmission rates associated with links that can be scheduled simultaneously. It uses column generation method to derive such feasible patterns in a computationally efficient way that is the solution to the optimization problem.

5.4 Power/Topology Control and Routing

A few research studies attempt to devise solutions for the routing and power control problem in conjunction. Neely et al. (2005b) present a formulation for dynamically optimizing power allocation and routing for time-varying channel characteristics and arrival rates. The capacity region of input rates are established, and a related joint routing and scheduling policy is presented that can stabilize the system with delay guarantees. The study of Kashyap et al. (2007a) considers joint topology control and routing problem for Free Space Optics (FSO) high speed mesh networks. FSO networks utilize high bandwidth, point-to-point narrow laser beam links. Such networks require topology control because FSO transceivers are expensive and actual links in the topology affect performance. They provide topology control and single/multipath routing algorithms (similar to wired optical networks) to choose efficient paths so that the FSO interface constraints are met, while traffic demands are also satisfied.

5.5 Routing and Channel Assignment

Finding routing paths with better channel diversity, or channel assignment for a given set of routing paths, is a challenging interdependent task. The study of Raniwala et al. (2004) provides one of the first centralized joint channel assignment and routing algorithms that takes into account estimated traffic demand and available channel/radio information. The algorithm recursively finds routing paths and corresponding channel assignments until the estimated traffic requirement is satisfied. Routing can be performed using a hop-count based shortest path algorithm, or load balancing multipath routing. The study of Raniwala and cker Chiueh (2005) extends the algorithm presented by Raniwala et al. (2004) for distributed design where nodes only have local information such as neighboring nodes and traffic load. A spanning tree rooted at the gateway is constructed for load-balancing routing, which uses hop-count, gateway link capacity or overall path capacity as metrics. Once the routing paths are found every node binds its neighbors with available radios (Neighbor-Interface Binding) and assigns channels to these interfaces (Interface-Channel Binding). The distributed algorithm presented requires local information only from $(k + 1)$-neighborhood (where k is ratio of interference range to transmission range).

An interference-aware channel assignment and QoS routing algorithm is presented by Tang et al. (2005). In the first phase, it performs topology control using channel assignment. In this phase, it finds minimum interference channels for links such that the topology is k-connected. In the second phase, an LP formulation is used to find feasible low-interference flow allocation on links. A new flow is admitted into the network only if such a flow allocation can be found. It provides a maximum bottleneck capacity path heuristic to ensure a single routing path between source and destination.

5.6 Scheduling and Channel Assignment

As discussed previously, partially overlapping channels (POCs) can be used to improve performance of WMNs, if used intelligently. The study of Mohsenian Rad and Wong (2007) performs scheduling and channel assignment of partially over-lapped channels as well as orthogonal channels, with assumption of some predefined routing mechanism. It introduces a channel overlapping matrix to systematically model the overlapping of the partially overlapped channels. Based on this, it presents a mutual interference model for all channels as an extension to SINR model for POCs. Using this it proves that interference range of receiver of a link depends on channel separation of that link to its neighboring links only. Considering this interference information of channels, it formulates channel assignment and scheduling as an LP formulation. The study of Liu et al. (2007) provides heuristics for channel allo-cation, and link scheduling for multiple partially overlapped channels with nodes having single-radio. It points out that channel sense mechanism of CSMA/CA MAC is not suitable for POCs as it waits for the medium to be free before transmitting. With POCs, transmission is still possible in overlapping channels and hence the proposed algorithm utilizes TDMA. It also proves that POC performs better with symmetric topologies because it achieves more spatial reuse in high density where more contentions are probable.

A novel approach is presented by Brzezinski et al. (2006) which partitions the network graph into subnetworks using *local pooling*. A static channel assignment algorithm is presented for partitioning the network such that each subnetwork has a large capacity region. Like a centralized approach, such a partitioned network can achieve 100% throughput when using distributed link scheduling.

5.7 Routing, Scheduling and Channel Assignment

Jointly optimizing routing, scheduling and channel assignment requires consider-ation of various parameters, and researchers have mainly investigated ILP based solutions for joint optimization. The study of Kodialam and Nandagopal (2003) first presented a solution to the joint routing and scheduling problem in single-radio multi-channel mesh under the assumption that there is a sufficient number of non-interfering channels available in the network. The study of Kodialam and Nandagopal (2005) extends the solution to multi-radio multi-channel mesh with limited number of available orthogonal channels. It provides an ILP formulation to maximize the total number of flows that can be supported by the network and meet node, channel, interference and flow constraints. It then tries to balance the flow load using dynamic or static channel assignment mechanisms while greedily scheduling the links simultaneously.

A similar LP formulation for joint channel assignment, routing and scheduling problem is presented by Alicherry et al. (2005). First, the algorithm tries to find paths achieving higher throughput with flow constraints and channel interference constraints. A channel allocation algorithm then modifies this solution based on available radios and number of assigned channels, to find a feasible channel assignment. Such modifications may require change of routes to maintain minimum interference. Such interference-free routes and channel assignments are then scheduled in conflict-free manner. In contrast to Kodialam and Nandagopal (2005), the approach of Alicherry et al. (2005) assumes that radios can not switch between channels during operation. An important departure of this problem was studied by Huang and Dutta (2006) which considers additive physical interference model (similar to geometric SINR (Goussevskaia et al. 2007)) instead of binary notion of interference. It presents two formulations for the problem—edge-based and node-based—and shows that the asymmetric node-based formulation is better suited for realistic additive interference model. It then presents a blossom-inequality based solution to solve a generalized matching problem. The study of Mishra et al. (2006) extends the LP formulation (Alicherry et al. 2005) of joint routing and channel assignment to use partially overlapped channels. With advancements in physical layer technologies, MIMO antennas are recently being adopted in 802.11n and 802.16 standards. Such MIMO links can send multiple data streams over its antenna elements independently. It can also eliminate interference with neighboring links if the total useful number of streams and interfering streams are less than the number of elements at the receiving antenna (Bhatia and Li 2007). A joint optimization problem for routing, scheduling and stream control using such MIMO links is also presented by Bhatia and Li (2007).

The study of Tarn and Tseng (2007) deals with joint routing, scheduling and channel assignment problem with TDMA-like MAC and dynamic channel assignment. First, it proves with a simple example that a multi-channel link layer and multi-path routing together can perform very well. It then proposes JMM (Joint multi-channel link layer and multi-path routing) protocol, which uses receiver-based channel assignment. In each slot of the super-frame, each node either sends or receives, resulting in interference-free transmission coordination. The number of transmit and receive slots in super-frame and its pattern is dynamically learned and changed depending on traffic requirements. The proposed forwarding strategy finds two disjoint paths from each node to the gateway while keeping broadcast overhead as low as possible. The proposed metric for finding routing paths captures link quality, channel diversity, and number of hops; to find minimum intra-flow and inter-flow interference routing paths. Overall, JMM achieves better performance by using multiple channels and paths together with timely coordinated transmissions.

5.8 Routing, Scheduling and Power Control

Routing, scheduling and power control decisions are highly interrelated and should be considered together for optimization. Cruz and Santhanam (2003) present one of the first solutions to this joint problem for multi-hop wireless networks. In the first phase, link scheduling and power control is performed with the objective of minimizing total power consumption. A feasible set of links and corresponding power levels are found with the constraints that each link has an average data rate no less than some given value and every node transmits at its peak power level in its assigned slot. To reduce the complexity of the solution with large number of links, hierarchical scheduling and power control is performed on clusters (Cruz and Santhanam 2003). These decisions are integrated in the second phase to determine routing paths. Routing facilitates the required data rates on each link based on source-destination traffic demand matrix. Similarly, Bhatia and Kodialam (2004) present a formulation for the joint optimization problem with the objective of minimizing power consumption with non-linear constraints of routing and scheduling. It provides a solution using a 3-approximation algorithm which yields a set of routes, a schedule and transmission power levels. The study of Kashyap et al. (2007b) presents a similar solution but there are no assumptions regarding prior knowledge of traffic matrix. Instead, it assumes that the traffic matrix always lies in a given polytope, which is derived using ingress and egress capacity of nodes.

Along similar lines, Li and Ephremides (2007) present a joint scheduling, power control and routing algorithm for TDMA-based wireless ad-hoc networks. In the first part, it is proved that performing scheduling together with power control yields better throughput and lower delay than obtained by scheduling separately. A centralized algorithm is presented where links are added to a feasible schedule, or removed based on scheduling rules (queue length at node, SINR constraint and disjoint endpoints); then per-link power control is performed. Running this algorithm without consideration of routing may cause congestion and bandwidth request to not always be satisfied. So, routing is integrated with scheduling and power control in the second part. The routing approach uses the Bellman-ford shortest path algorithm on a metric that captures the effect of traffic congestion and link conflicts. Simulation results show that this joint approach can yield better network performance.

An interesting trade-off of *larger-range lesser-hops* and *shorter-range more-hops* is pointed out by Pathak et al. (2008). It shows that if high power transmissions are used, it gives rise to long, high-interference links. Such links preclude scheduling many other links, but the data reaches the destination in fewer hops, with lower delay. Instead, if low power transmissions are used, the data reaches the destination via many hops, but all such shorter links can be scheduled with a larger number of other links. It is an open question whether any of these two mechanisms perform better in terms of throughput and delay. The study of Pathak et al. (2008) introduces the concept of *loner links*, the links which cannot be scheduled with any other link in the network due to their high interference characteristics. Traffic on such links should be re-routed via shorter low interference links. Analytical characterization of loner links is also presented for square or circular network areas.

5.9 Designing Using Sustainable Energy

Wireless mesh networks have the ability to facilitate a tetherless deployment since the mesh nodes are already connected to the gateways through wireless links. It is important to note that as opposed to many low-power systems such as sensor networks, mesh networks can not operate using low-power batteries. This is because there are two fundamental requirements of a mesh network that are different from other low-power networks:(i) mesh nodes are required to provide larger coverage area, and (ii) mesh nodes should not only provide continuous access to clients but should also relay traffic of other mesh nodes. Due to these mesh nodes can not tolerate disruption in its service which might be possible in sensor networks using sleep/wake up cycles.

The objective of tetherless deployment of mesh nodes with the objective of uninterrupted service are at odds if the mesh nodes are not powered using an uninterrupted power supply. With recent initiatives and resultant breakthroughs in sustainable energy sources, it has in fact become possible to deploy and operate mesh nodes using renewable energy sources such solar power or wind power. Apart from obvious advantages of using clean energy, this allows a tetherless deployment of mesh nodes where it is in fact possible to reformat the network without significant efforts.

The potential of utilizing renewable energy sources for mesh nodes brings along numerous challenges that needs to be addressed. A disadvantage of renewable energy in current state of technology is that the availability of energy demonstrates significant spatial and temporal variation. This is a challenge for mesh networks because mesh nodes require uninterrupted power supply. Specifically, when mesh nodes are operating using such energy sources, their outage becomes more probable mainly due to unavailability of energy. Also, it is possible that a mesh node gathers data from client or other mesh nodes but can not forward it further towards the gateway due to unavailability of power to transmit the data. This results into failure of service which is not tolerable especially when data flows are inelastic in nature.

To address this issue of low availability of energy but high required availability of mesh nodes, various design approaches have been suggested. At large, these efforts can be divided into two high level design goals:

1. Understanding and modeling energy availability
2. Designing and deploying mesh nodes using energy availability analysis

Both the objectives are discussed in the following sections.

5.9.1 Understanding and Modeling Energy Availability

Since it is not desirable for mesh nodes to drop client data or buffer it indefinitely, novel approaches are needed which can design the network accordingly. A necessary component for such approaches is that of understanding the energy availability from renewable sources, and modeling it to estimate its future. Two commonly used renewable energy sources are discussed here—solar and wind. Energy availability in both types largely depend on environmental conditions. These conditions can modeled using techniques that depend on weather forecast, conditions of immediate past etc. Various such methods are described below.

5.9.1.1 Past Predicts the Future (PPF) of Energy Availability

PPF approach is based on the idea that recent past of climate conditions can be used in order to predict their value in future. These climate conditions can be then used to predict the energy availability of near future. The strength of the approach lies in to the fact that it does not require any external information such as weather forecast to estimate the energy availability. As stated by Sharma et al. (2010),

> PPF states that value of the metric representing a weather condition in next T time units will match exactly to its observed value in last t time units.

Due to its simplicity, PPF has been largely used in literature for modeling solar energy availability. The drawback of PPF scheme is that it reacts very slowly to sudden changes in climate conditions. Studies like those of Fan et al. (2008), Kansal et al. (2004), Moser et al. (2007) and Noh et al. (2009) show that it is a good estimator for very short (in scale of minutes) and very long (in scale of months) future, but it fails to capture the changes in a more reasonable time-scale of future.

To address this, Noh et al. (2009) have suggested the usage of moving average. Formally, let e_{t_i} be the energy generated in slot t_i, now expected energy available during slot t_i can be updated using

$$E[e_{t_i}] = (1 - \theta)E[e_{t_i}] + \theta e_{t_i} \tag{5.3}$$

here, $0 < \theta < 1$ is a controllable parameter which can be used to dictate how much consideration should be given to historical data as opposed to data observed in the current time. Using this method, PPF can be extended for better accommodation of short-term changes.

For much more improved consideration of temporary conditions in predicting energy availability, Noh et al. (2009) make a different contribution. PPF scheme presented by Noh et al. (2009) (referred as MPPF—Modular PPF) maintains a ratio σ of actual amount of energy generated and its expected value for previous time

Fig. 5.1 Diurnal variation of generated solar power: the peak occurs during daylight hours, and is correlated with daytime temperature

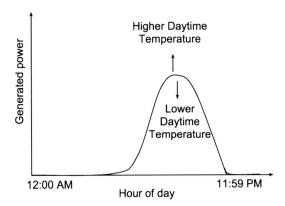

slot, hence $\sigma = \dfrac{e_{t_{i-1}}}{E\left[e_{t_{i-1}}\right]}$. The ratio σ is representative of more current past, and $\sigma < 1$ indicates that the actual generated energy was not well predicated using the expectation, hence the immediate past may have observed some sudden climate changes.

Assuming that a day is divided into n time slots, energy generated in any time slot *imodulon* can be now estimated using e^* where

$$e^*_{imodulon} = \sigma E[e_{imodulon}] \tag{5.4}$$

Note that it is in fact possible to intelligently vary the value of σ depending on specific information (such as time of day etc.).

It is also worth noting that PPF schemes are only capable of predicting solar generated power, and are not applicable to wind power generation. This is mainly because wind power generation does not display diurnal behavior, and it is indeed dependent on large number of other climate variables.

5.9.1.2 Leveraging Weather Forecasts

One limitation of PPF methods is that they can only comprehend local climate information for prediction. There is a large number of global forecast services that can provide more accurate future climate information. Forecast-based energy availability prediction (FEAP) methods incorporate forecast information to estimate energy availability more accurately. Sharma et al. (2010) made the first contribution towards developing and validating a FEAP method. In most cases, better estimation of energy availability allows mesh nodes to plan the transmission of elastic and inelastic data flows.

One significant advantage of FEAP methods over PPF methods is that it can be applied to wind power generation. Sharma et al. (2010) used actual solar panel and wind turbine testbed to measure the amount of energy generated over 12 days

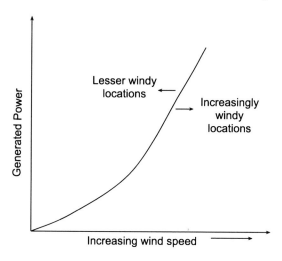

Fig. 5.2 Generated wind power increases sharply with wind speed; the variation is less pronounced for turbulent and more pronounced for steady wind locations

using solar and wind sources. They reported that the per day variation observed in generated energy is much higher for wind power since it does not have the diurnal property. It is also worth noting that there is substantial amount of variation observed in generated solar energy across 12 days, mainly due to cloudy weather and related climate changes. PPF methods can not estimate energy availability in such day-by-day varying conditions, and FEAP methods are useful in such cases. Sharma et al. (2010) showed that PPF methods are unable to accurately predict the future in medium-length scale from 3 h to 1 week.

To further understand how forecast can be used for prediction, Sharma et al. (2010) studied how generated solar power varies in a day. The general nature of their findings are shown in Fig. 5.1. They observed that this expected pattern of maximum solar power generated can be mapped to a quadratic function of time,

$$Maximum - Power = a * (Time + b)^2 + c \qquad (5.5)$$

Note that the values of a, b and c are dependent on the time of the year, and Sharma et al. (2010) described their values for each month of the year. The values are obtained using curve fitting procedure. Now, this information can be treated as an optimistic estimate which can be brought down to a realistic estimate using forecast information. They advance an idea that if the sky condition forecast shows that there will be a cloud cover of N % then the measured solar power will be (100−N) % of the maximum possible power. Based on this, estimated solar power can be now found using

$$Power = Maximum - Power * (1 - SkyCondition) \qquad (5.6)$$

Similarly, for wind power estimation model is derived by Sharma et al. (2010) using curve fitting. Figure 5.2 shows in general how generated wind power is related to wind speed. The wind speed information is available through forecast service which can be used to estimate generated wind power. Using the curve fitting, it is shown that wind power is a cubic function of wind speed.

$$WindPower = 0.0179 * (WindSpeed)^3 - 3.4 \qquad (5.7)$$

Apart from the climate factors studies above, Sharma et al. (2011) studied other factors which show some correlation with solar power and wind power generation. They found that wind speed, dew point and temperature show only moderate correlation with solar power. On the other hand, sky cover, relative humidity and precipitation chances show a high inverse correlation with solar power. Since these information are available through forecast services, a regression analysis can be used to estimate the solar power using their values, as also presented by Sharma et al. (2011).

5.9.1.3 Markovian Model for Solar Radiation

When there is a large amount of historical data available, Markov model can be used for modeling solar radiation and resultant solar power. As shown by Niyato et al. (2007), the model can be extended to accommodate cloud thickness, wind speed and other climate variables.

In such markovian model of solar radiation, let R be an array representing solar radiation states $R = \{R_1, R_2, ..., R_N\}$, and P be a N × N probability matrix where each element P_{R_i, R_j} is the probability of solar radiation state transition. The elements of matrix P can be derived using historical data of the site where mesh network will be deployed.

It was shown by Niyato et al. (2007) that P can also be derived using from other probabilities such as probability of a certain wind speed, certain cloud size, and their duration. The probabilities can be based on known probability distribution such exponential distribution. Various characteristics of the markov chain such as steady-state probability etc. can be found and used in energy availability analysis.

One limitation of this model is that a large amount of historical data is necessary in order to calculate the transition probabilities which makes the model less reactive to short-term changes.

When a years worth of historical data is available, it is in fact possible to take expectation of solar radiation on hourly bases, and use it directly for estimating energy availability. An obvious limitation of this is that such a method can not consider short-term changes in climate.

A sample diagram of a solar mesh node is shown in Fig. 5.3 following Farbod and Todd (2007). The charge controller is responsible for making sure that the battery is not over or undercharged at any time.

Fig. 5.3 Design of a solar
mesh node: a charge controller
uses photovoltaic current to
charge a battery, inverted
power powers mesh node

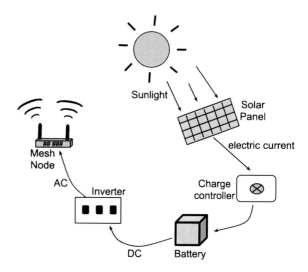

5.9.2 Design and Deployment of Renewable Energy Mesh

With the methods described above, a network designer can model the energy availability, and use the modeling analysis for informed design of mesh networks. As
opposed to the design problems discussed in Chaps () and (), design of renewable
energy mesh adds few interesting challenges. These challenges are discussed next.

5.9.2.1 Planning and Deployment

As mentioned before, mesh nodes can not tolerate service disruption due to unavailability of energy. This requires that the mesh networks is planned in such a way that
the outage of mesh nodes can be gracefully accommodated.

Node placement: For all potential locations where mesh nodes can be deployed,
their energy availability and outage probability should be calculated. This together
with the expected traffic demand at various locations should be considered to determine the locations where mesh nodes can be placed. Note that it is likely that outage
probability of mesh nodes can not meet the demand at many locations mainly due to
the use of renewable energy sources.

To avoid this issue of meeting demand with given outage probabilities, redundant
nodes should be deployed. Since multiple nodes now provide coverage to the same
area, it is in fact possible to load balance between them as per their energy availability.

Hybrid Deployment: In the cases where traffic demand at various locations can
not be satisfied using renewable energy sources, mesh nodes with fixed power source
can be used. This results into a hybrid deployment where low to moderate traffic
demand locations are served using mesh nodes with renewable sources and high

traffic demand locations are served using mesh nodes with uninterrupted power sources.

Network Expansion: One fundamental advantage of mesh network is network expansion without incurring a large cost. Network expansion should take place in case where additional mesh nodes are added due to traffic growth or increased coverage requirement. When performing network expansion, it is necessary that factors such as outage probability and traffic demand are accommodate in design to provide uninterrupted service.

Some research has been performed to address these problems of node placement (Vargas et al. 2007; Farbod and Todd 2007; Zefreh et al. 2010), hybrid deployment (Sayegh et al. 2008) and network expansion (Badawy et al. 2009b).

5.9.2.2 Protocol-Level Design Problems

Since the traffic in mesh network is relayed to gateways in a multi-hop manner, it is necessary to consider the energy expenditure incurred by relaying. This becomes a challenge especially for renewable energy mesh nodes since the outage probability of nodes performing more relaying increases.

The problem can be circumvent by utilizing energy aware routing protocol which allows dynamic balancing of energy resources while forwarding the data. The problem is similar to that of energy-efficient routing protocols for sensor networks for which there exists a good level of understanding in literature. Such protocols can be extended to accommodate newly generated energy at various nodes which can be predicated beforehand, and routing can be performed accordingly. As an example, a mesh node that has low residual battery charge can be selected for relay if it is expected to generate sufficient amount of energy in near future. Such research problems are open and future research efforts should be directed towards addressing them. Some existing research has addressed these issues by Badawy et al. (2008a,b, 2009a, 2010), Farbod and Todd (2007), Sayegh and Todd (2007).

5.10 Back-Pressure: A Framework for Maximizing Network Utility

The back-pressure scheduling/routing policy first proposed by Tassiulas and Ephremides (1990) has recently shown a great potential for solving a number of issues in wireless multi-hop networks. The central idea of back-pressure scheduling policy is that contention among the links should be resolved by scheduling the link which has the largest product of queue differential backlog between its endpoints and transmission rate at which the link can be served. In a perfectly time-slotted medium access mechanism such as TDMA, this will result into optimal throughput of flows while guaranteeing queue stability (ingress traffic to a queue never exceeds its egress traffic). The utility maximization framework initially proposed by Kelly et al. (1998) shows that injection rates of flows should be chosen such that aggregate utility of the flows is maximized. Here the utility of flow represents a desirable effect on the network achieved by a particular rate of the flow. It was shown by Chen et al. (2006), Eryilmaz and Srikant (2006), Lin and Shroff (2004) and Neely et al. (2005a) that back-pressure scheduling and utility based rate control together can solve the global problem of network utility maximization.

We first provide a brief overview of the back-pressure framework. The back-pressure framework consists of two parts: a link scheduling strategy based on back-pressure and a rate control module for network utility maximization.

Back-pressure scheduling: Back-pressure based link scheduling policy was originally proposed by Tassiulas and Ephremides (1990). Let's represent the network under consideration using a graph $G = (V, E)$. Let f denote a traffic flow from source node $s(f)$ to destination node $d(f)$ and F be the set of all flows currently active in the network. Let $l(u, v) \in E$ denote a link between node u and v, and $\gamma_{l(u,v)}$ be the transmission rate of link $l(u, v)$. Transmission rates of all links in E are presented by $\Gamma = \{\gamma_{l(u,v)}, l(u, v) \in E\}$. Let χ be the set of all possible combinations of rates at which links can operate.

Let us assume only for now that the transmission time is divided into equal sized time slots. Every node $u \in V$ maintains a separate queue for destinations of all flows in F. These queues at each node are also known as per destination queues (PDQs). Packets received by u for a flow f destined to $d(f)$ is stored in the queue $Q_u^{d(f)}$ until further forwarding decisions are made. Let $|Q_u^{d(f)}(t)|$ denote the size of the PDQ maintained at node u for destination $d(f)$ at time t. Every node shares its PDQ length information with all its neighbors at the beginning of time slot t. Every node then calculates its differential backlog as compared to its neighbors for every flow destination. That is, a node u calculates $D_{l(u,v)}^{d(f)}(t) = |Q_u^{d(f)}(t)| - |Q_v^{d(f)}(t)|$ for all $l(u, v) \in E$ and all $f \in F$. Now, for every link $l(u, v) \in E$, let

$$\Delta_{l(u,v)}(t) = \max_{f \in F} \left(D_{l(u,v)}^{d(f)}(t) \right) \qquad (5.8)$$

Back-pressure scheduling suggests that Γ at time t should be chosen such that

$$\Gamma(t) = \max_{\Gamma \in \chi} \sum_{l(u,v) \in E} \left(\gamma_{l(u,v)} \Delta_{l(u,v)}(t) \right) \qquad (5.9)$$

It was proved by Tassiulas and Ephremides (1990) that a routing/scheduling policy that can achieve a solution of Eq. 5.9 is throughput optimal. This means that it stabilizes the queues at every node while supporting the largest possible capacity region. A capacity region (C) of a network is defined as the set of all flow rates which are supportable by the network. Due to link interference constraints, the above mentioned problem is proven to be NP-hard in wireless networks. Also, its distributed implementation imposes many more challenges in design of a routing/scheduling scheme for wireless networks.

Rate control: The back-pressure scheduler determines the transmission rates at which packets are served at links so that the queue sizes at nodes remain bounded while maximizing the achievable throughput. The other part of the back-pressure framework performs the rate control of the flows. The flow controller determines the rate at which flows can inject the packets in the network. In back-pressure framework, the flow controller determines the input rates of flows based on the state of queues at each intermediate node. The PDQ information is utilized by the flow controller to adjust the flow input rates in a way that a desirable network-wide objective is optimized. In a seminal work, Kelly et al. (1998) showed that such flow/congestion control can be viewed as primal-dual algorithm for the solution of network utility maximization problem. This was further elaborated by Akyol et al. (2008) in context of the back-pressure framework which we describe next.

Suppose that each flow f from $s(f)$ to $d(f)$ has a utility function associated with it. This utility function $U_f(x_f)$ is a function of the rate x_f of the flow f. Let us represent input rates of all flows using $\psi = \{x_f, f \in F\}$. The utility maximization problem suggests that the flow rates should be chosen such that their aggregate utility is maximized, that is:

$$\max_{\psi \in C} \sum_{f \in F} U_f(x_f) \tag{5.10}$$

A flow controller that can maximize the aggregate utility was presented by Akyol et al. (2008). It suggests that in each time slot t, source $s(f)$ of a flow f injects a packet in the network if and only if

$$U'_f(x_f(t)) - \beta Q^{s(f)}_{d(f)}(t) > 0, \tag{5.11}$$

where U'_f is the first derivative of utility function of the flow and β is a small constant. The above condition interprets to the fact that a packet should be injected into the network by a flow only if the eventual utility benefit of the insertion is larger than a constant times the size of source node PDQ. That is when intermediate nodes of a flow are sufficiently back-logged, the source node pushes more packets into the flow at a slower rate. The back-pressure scheduler ensures that queue backlog status of intermediate nodes is reflected back at the source node which then performs the rate control to accordingly change the flow input rate.

This way, the back-pressure based link scheduling controls which links should be transmitting at what rate, and the rate control module manages the rates at which the packets are injected by the flows. Both together can solve the network-wide utility maximization problem to yield a throughout-optimal solution.

5.10.1 Using CSMA/CA in Back-Pressure Framework

The fundamental challenge with back-pressure framework is that solution of the underlying scheduling strategy is NP-hard (Georgiadis et al. 2006). Also, since it was proposed for a centralized, synchronized and time-slotted system, a distributed implementation which can achieve even a closer approximation is very difficult to develop. Recently, Akyol et al. (2008), Warrier et al. (2009) have attempted to incorporate back-pressure based scheduling in random medium access protocols such as CSMA/CA. These protocols try to approximate the performance of the ideal back-pressure scheduler by prioritizing the frame transmissions according to differential backlogs of queues. Here, every node in the network maintains a per destination queue (PDQ) and the packets destined to a particular destination are stored in the PDQ of that destination until further forwarding decisions are made. Now, nodes share their PDQ information with their neighbors, and this information is utilized by every node to calculate differential backlogs of its PDQs. The differential backlog of a PDQ at a node is equal to the size of the PDQ minus the size of the PDQ of its upstream neighbor towards the destination. To emulate the back-pressure scheduling, packets of the PDQ which has the highest differential backlog (highest back-pressure) in the neighborhood are given the higher chances for transmission. This way, the likelihood that packets are transmitted from a particular PDQ at a node is proportional to its differential backlog compared to the differential backlogs of PDQs of all nodes in the neighborhood. This prioritization quickly moves the traffic from long back-logged queues to shorter queues achieving an improved throughput and a better overall stability of queues.

In back-pressure scheduling, every node maintains a PDQ for every flow passing through the node. Note that we assume a fixed routing policy in this work where routes are calculated in advance for every flow and packets are forwarded on these routes only. This is different from an ideal back-pressure strategy where the routing is adaptive, and forwarding decisions are made on packet by packet basis. In the ideal scheme, every neighbor of a node depending on its size of PDQ is a potential candidate for forwarding the packet (Ying et al. 2010). Since our objective is to explore packet aggregation with back-pressure strategy, we restrict our focus to fixed routing and leave the adaptive routing extension to future work.

Due to fixed routing, every node has a unique upstream neighbor for every flow passing through it. Now, the differential backlog at a node u for flow f can be presented as $D_u^{d(f)}(t) = |Q_u^{d(f)}(t)| - |Q_v^{d(f)}(t)|$, where v is next hop neighbor of node u for flow f. Each node maintains following information (Warrier et al. 2009; Akyol et al. 2008) in order to correctly execute the joint back-pressure and aggregation scheme:

1. Per destination Queue (PDQ)—a separate queue for each destination of flows passing through the node. Every packet (generated at the node or received from downstream neighbor) is stored in the PDQ of corresponding destination until further forwarding decisions are made.
2. Urgency Weight—every PDQ has an urgency weight associated with it. The urgency weight is the differential backlog which equals to the backlog of the

PDQ subtracted by the backlog of the PDQ on the next hop neighbor towards the destination.

3. Urgency Weight State (UW state)—along with maintaining PDQs and their corresponding urgency weights, every node knows the PDQ ID, node ID and urgency weight of the PDQ which has the maximum urgency weight in the neighborhood. The same is also maintained for the minimum urgency weight PDQ of the neighborhood. This information determines a node's UW state with respect to its neighbors. To implement this, every node shares its PDQ information with its neighbors by using separate messages or piggybacking techniques.

4. Source List—source node of every flow which generates packets first stores the packets in a per destination source list. Once the rate control is performed then only the packets are added to the corresponding PDQ at the node. Different from the PDQs which are maintained at all nodes, per flow source lists are only maintained at the source nodes of the flows.

In a back-pressure scheduling implementation without aggregation (Akyol et al. 2008; Warrier et al. 2009), when a node sends out a frame from a PDQ, it first determines its MAC layer priority. This is described in *Determine-MAC-priority* function of Algorithm 1. The MAC priority is essentially determined by comparing the urgency weight of the PDQ to the UW state of the node. In essence, this compares the differential backlog of the PDQ with the differential backlog of other PDQs in the neighborhood and assigns it a MAC priority based on it. Packets of the PDQ which has the highest urgency weight in the neighborhood are assigned the highest MAC layer priority for faster transmissions. Here, it is assumed that MAC layer is able to provide differentiated levels of service (such as 4 distinct MAC layer priorities). Packet sent out with higher priority has higher chances of accessing the channel as compared to the packets sent out at lower priorities. CSMA/CA MACs like 802.11e provides such differentiated service levels which we also use in our simulations. This prioritization of transmissions tries to serve the longer queues with higher service rates which is the ultimate objective as described in Eq. 5.9.

Warrier et al. (2009) developed a rate control protocol (called DiffQ) using MAC priorities as described above. Network utility is measured as the sum of logarithm of flow throughput rates ($\sum_{f \in F} log(x_f)$). For throughput fairness of flows, it uses Jain's index (Chiu and Jain 1989) which is defines as—

$$\text{Fairness index} = \frac{(\sum_{f=1}^{m} x_f)^2}{m \sum_{f=1}^{m} x_f^2} \qquad (5.12)$$

where m is total number of flows in the network and x_f is the throughput of every flow. Warrier et al. (2009) compare the performance of DiffQ to TCP-FeW (Kanth et al. 2002) and TCP-SACK. using an actual testbed. They observed that DiffQ can achieve far superior fairness index compared to the native schemes. DiffQ achieves the same order of, and in many cases a slightly lower, throughput. However, DiffQ improves overall network utilization while maintaining fairness across flows.

5.10.1.1 Packet Aggregation

The idea of using different MAC layer priorities for PDQs with different back-pressure works well in imitating the ideal back-pressure scheduling over CSMA/CA. Even though this enables a distributed implementation, it does not guarantee that only the frames from maximum back-logged queue are being transmitted at any given time. This is because MAC priorities themselves are implemented by modifying the size of the contention window from which the back-off times are randomly chosen. A higher MAC priority frame uses a shorter random back-off time before being transmitted, and it is possible that frames from queue with lower differential backlog get transmitted before frames from queue with higher differential backlog. This problem of not serving the most back-logged queue with the highest possible rate is further aggravated by the fact that most of the packets in last mile networks such as wireless mesh networks are very small in size. Jain et al. (2003); Na et al. (2006); Tang and Baker (2000) show that application layer traffic and TCP acknowledgments account for more than 70–80 % of total traffic where packet sizes are lesser than 100 bytes. With such small packet sizes, time required for medium access and header/trailer overheads adversely affect the overall performance. In case of prioritized frame transmissions, this further reduces the chances that the most back-logged queue is obtaining the maximum possible medium access. Even if the packets are being transmitted from that queue, additional time required for medium access before transmitting every small packet accumulates to a large waste of available air time. Both these factors reduce the rates at which back-logged queues are served which in turn reduces the throughput and network utilization.

Deuskar et al. (2012) propose utilizing packet aggregation in the back-pressure framework. The objective is to increase the throughput while maintaining the inherent properties and benefits of back-pressure scheduler. The central idea is to take advantage of packet aggregation for improving the rates at which the back-logged queues are served. Different from other packet aggregation schemes (Ganguly et al. 2006; Jain et al. 2003; Kliazovich and Granelli 2008; Raghavendra et al. 2006; Riggio et al. 2008; Zhai and Fang 2005), the presented strategy utilizes the back-pressure invariants to guide when and how much aggregation is performed. Specifically, when a queue has the highest differential backlog in the neighborhood, instead of transmitting all its head of line packets one by one, the proposed scheme aggregates the first few packets in a large MAC frame which is then transmitted with the highest MAC priority. The number of packets that are aggregated depends on maximum allowable MAC frame size and the differential backlog of the queue. The proposed scheme attempts to reduce the size of the back-logged queues more aggressively which results into their faster service rates. Since the back-pressure policy is utilized to perform the aggregation, the presented scheme preserves the network utility and fairness advantages of the back-pressure policy while yielding an improved throughput performance. This leads to a closer approximation to the optimal solution of back-pressure problem. Apart from throughput improvements, aggregation based back-pressure solution reduces the end-to-end delay of packets dramatically which is especially important in delay-constrained applications.

Packet aggregation: The service rates of back-logged queues can be further increased if more and more packets are transmitted from the queues with higher urgency weights. Note that the above implementation of back-pressure strategy without aggregation tries to maximize the chances that medium is occupied by the packets transmitted from the longer queues, but it does not guarantee to do so. Since a large number of packets are very small in size and medium access has to be initiated before every transmission, it is not guaranteed that the packets of the longest queue in a neighborhood receives the highest possible service rate from the medium. Instead if the packets of the longest queues are aggregated before their medium access and eventual transmission, more packets can be sent out from the backlogged queue in each medium access. This eliminates the additional time required for medium access by every small packet and increases the total air time utilization by longer back-logged queues. This increases the service rates of back-logged queues which in turn maximizes the objective function presented in Eq. 5.9. It is obvious that back-pressure with aggregation is also an approximation of the optimal solution to Eq. 5.9 but is closes the gap further towards the optimal solution by improving on any currently available scheme.

Algorithm 1 Packet aggregation + back-pressure scheduling

$maxAggSize$ ←Maximum allowable size of a MAC frame; $size$ ← 0,
q ←PDQ with the highest urgency weight,
repeat
   Dequeue the HOL packet p from PDQ q,
   Add p to the packet aggregate,
   $size := size + \text{sizeof}(p)$,
until $size \leq maxAggSize$ **AND** PDQ q has the highest urgency weight in the neighborhood
Determine MAC priority of the aggregated packet using Determine-MAC-priority(q),
Transmit the packet with the MAC priority,
Update the size and urgency weight of PDQ q.
Determine-MAC-priority(q)
   $numLevels$ ← Number of MAC priority levels,
   max ←Maximum urgency weight in the neighborhood,
   min ←Minimum urgency weight in the neighborhood,
   urg ←Urgency weight of the PDQ q,
   $macPrioLevel := \frac{urg-min}{max-min} * numLevels$,
return $macPrioLevel$

Aggregation based PDQ scheduler at every node de-queues the head of line packets from PDQ with the highest urgency weight. The node then performs the check whether the PDQ is in fact the highest urgency weight PDQ in the entire neighborhood. If so, it de-queues more packets from the PDQ. It continues to do so until the PDQ retains the highest urgency weight in the neighborhood or until sum of the size of all the de-queued packets is less than the maximum allowable MAC frame size. All the de-queued IP packets are then bundled into a large MAC frame of aggregated packets. The MAC frame is then assigned a MAC priority depending on the current UW state of the node. Since the PDQ from which the packets were de-queued had the highest urgency weight in the neighborhood, it is also likely that the MAC priority of the frame of aggregated packets will be the highest too. Because of aggregation the total time taken to transmit these set of packets will be reduced as compared to their individual transmissions. This increases the service rate of the PDQ with highest urgency weight in the neighborhood yielding a closer approximation of Eq. 5.9. The complete procedure of aggregation and back-pressure scheduling is described in Algorithm 1.

Rate control: While the back-pressure scheduling and aggregation is performed at every node of the network, source node of every flow performs the rate control. The function of this rate control policy is to determine at what rate the packets should be injected in the network. The objective of such a flow control is to maximize the utility (or benefit) associated with each flow. Here, whether a packet is injected into a flow or not depends on the utility function of the flow and the size of PDQ at the source node for the flow. Specifically, a flow f injects a packet into the network as long as $U'(x_f) > \beta Q_{d(f)}^{s(f)}(t)$, where $U'(x_f)$ is the first derivative of the utility function U_f. We use $U_f(x_f) = log(x_f)$ as the utility function as described by Akyol et al. (2008). β is a small constant whose value is set to 10^{-6}.

This means that the source node $s(f)$ of a flow f first checks whether $U'(x_f) > \beta Q_{d(f)}^{s(f)}(t)$ condition holds. If the condition holds true, the source node inserts the packet in its PDQ for flow f. On the other hand, if the condition does not hold true, the source node inserts the packets in the source list of the flow. The source list acts as a tentative storage where packets are buffered until rate control permits them to be added in corresponding PDQ. When back-pressure scheduler transmits packets from the PDQ, packets are added in the PDQ from the source list using the rate control condition. At all times, any addition or removal from PDQ is followed by recalculation of urgency weight of the PDQ. If the source node PDQ reflects that intermediate nodes are sufficiently back-logged, it slows down the rate at which the packets are added from the source list to the PDQ.

Figure 5.4 (Deuskar et al. 2012) shows the aggregate throughput achieved with both back-pressure and aggregation based back-pressure schemes. The aggregation based strategy improves the throughput significantly mainly because the backlogged queues are served at faster rates when aggregation is utilized. Without aggregation, every packet of backlogged queues requires additional time for medium access. Also, overhead of MAC layer headers for each frame without aggregation also accounts for a large wastage of bandwidth. Both these issues are well addressed by the utilizing aggregation along with back-pressure policy.

As it was described by Deuskar et al. (2012) before, the back-pressure strategy was mainly devised to address low network utilization and unfairness issues in multi-hop wireless networks. Even though aggregation with back-pressure scheduling increases the network throughput, it is necessary to verify that it does not do so by penalizing network utilization or fairness. Figure 5.5 shows the network utility of back-pressure scheme with and without aggregation. It can be observed that in fact aggregation can achieve same or sometimes more network utilization than back-pressure scheme without aggregation. This is mainly due to higher service rates of back-logged queues

Fig. 5.4 Aggregation consistently improves throughput
over PDQs only

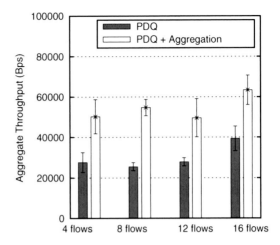

Fig. 5.5 Similar and somewhat better network utility is
obtained with aggregation

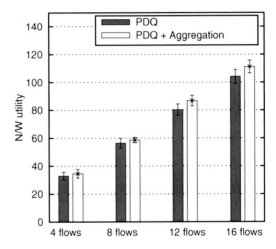

which in turn allows rate control to insert more and more packets in the flow, yielding
an improved utilization. Similarly, Fig. 5.6 shows the fairness index calculated for
both schemes. It is observed that aggregation also increases the throughput fairness
in most cases. This is attributed to the fact that aggregation increases the air time
utilization of back-logged queues more aggressively which results into improved
balance among flow rates.

5.10.1.2 Optimal CSMA

A limitation of above mentioned CSMA/CA based utility maximization is that it
requires message passing among the nodes in order to calibrate the urgency weight.

Fig. 5.6 Fairness is not
seriously impaired by aggregation, and is sometimes
improved

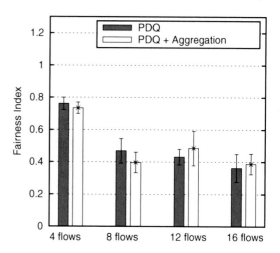

Also, since the method is developed using only 4 discrete priority levels, it does
not guarantee an optimal solution to the utility maximization problem. In designing
a purely distributed algorithm which does not require any message passing but is
still provably optimal is a challenging problem. The problem has been addressed in
various recent research efforts (Durvy and Thiran 2006; Jiang and Walrand 2010;
Jiang and Liew 2008; Lee et al. 2010; Liu et al. 2010; Marbach and Eryilmaz 2008;
Ni et al. 2012; Rajagopalan and Shah 2008; Rajagopalan et al. 2009) that have design
such a CSMA variant that is throughput optimal and maximizes the network utility.
The variant is referred as Optimal CSMA or Q-CSMA, and is described next.

For O-CSMA, the time is assumed to be divided into slots. We borrow the notations
from Nardelli et al. (2011), and presented the algorithm of Nardelli et al. (2011). Let
the mean of random back-off counter of CSMA denoted by $1/\lambda$, and mean of random
transmission durations be μ. $CSMA_l(\lambda_l, \mu_l)$ is used to denote when a link l uses O-
CSMA with λ_l and μ_l parameters. Let $q_l[t]$ denote a virtual queue maintained by the
link l at time t.

1. During frame t, the transmitting node if link l executes $CSMA_l(\lambda_l, \mu_l)$, and
   records the number of packets $S_l[t]$ that were serviced during the time frame t
   on link l.
2. At the end of frame t and before starting frame $t + 1$, the node updates its virtual
   queue as follows:

$$q_l[t+1] = \left[q_l[t] + \frac{b[t]}{W'(q+l[t])} \left(U'_{-1} \left(\frac{W(q_l[t])}{V} \right) - S_l[t] \right) \right]_{q_{min}}^{q_{max}} \quad (5.13)$$

3. Set $\lambda_l[t + 1]$ and $\mu_l[t + 1]$ so that their product equals $exp\{Wq_l[t + 1]\}$.

Here, b is a step size function that ensures the convergence of the algorithm. W
is a strictly increasing and continuously differentiable function. Other parameters

V, q_{min}, q_{max} are positive parameters, while $[x]_a^b = min(b, max(a, x))$. The rate control algorithm remains the same as described above. The O-CSMA ensures that links not able to get service from the medium are likely to more aggressively try and occupy medium, but the advantage of the method is that it does so in distributed matter without requiring any message passing.

The O-CSMA scheme was analyzed using experiments by Nardelli et al. (2011). The purpose of the study is to identify limitations of O-CSMA when applied to realistic traffic scenarios. Few of the fundamental assumptions of O-CSMA such perfect carrier sensing, bi-directionality of links and interference, and absence of coordination between MAC layer backlog and upper layer protocols make it difficult to be applicable to real-world scenarios. Nardelli et al. (2011) used a few representative topologies in order to detect the limitations of O-CSMA, and the findings are summarized below.

1. Since O-CSMA design does not consider hidden terminals and resultant collisions, it performs poorly in presence of collisions. This is because unaware of causes of collisions, O-CSMA increases transmission aggressiveness which in turn increases the collisions. This results into poor throughput even with use of RTS/CTS.
2. The current design of O-CSMA is insufficient to mitigate the information asymmetry problem (where out of two flows, one flow interferes with other but not vice versa). There is some improvement in terms of achievable throughput but it without the use of RTS/CTS the advantages remain limited.
3. Flow-in-the-middle problem (where two non-interfering flows interfere with a third flow in middle) can be mitigated by O-CSMA. The results are similar to those of Warrier et al. (2009) as described before.
4. In case of channel asymmetry problem (where single transmitting node sends data to two neighbors over two links of varying loss rate), O-CSMA can indeed achieve comparable throughput irrespective of link capacities.

To address the issue of performance of O-CSMA in presence of imperfect carrier sense, Eryilmaz and Srikant (2006) showed that O-CSMA can in fact achieve an arbitrary fraction of capacity region if certain parameters and access probabilities are set intelligently. Specifically, it shows that $(1 - \gamma a)^\delta$ throughput can be achieved where γ is the probability by which the per-link carrier sensing function fails to detect an ongoing transmission, a is the access probability and δ maximum degree of interference conflict graph. This way, setting a appropriately indeed achieves a throughput that is arbitrarily close to one.

5.10.2 Joint Routing, Scheduling and Rate Control Using Back-Pressure Framework

The back-pressure framework of utility maximization comprise of a routing module that can forward data on multiple paths, a rate control module that controls the injection rate, and a scheduling module which maximizes the medium occupancy of back-logged queues. The approaches discussed till now did not consider a multi-path routing module and assumes that routing is pre-determined. A large body of literature (Eryilmaz and Srikant 2005, 2006; Georgiadis et al. 2006; Lin and Shroff 2004; Lin et al. 2006; Neely 2006; Neely et al. 2005a; Neely and Urgaonkar 2009; Stolyar 2005; Yeh and Berry 2007a,b; Ying et al. 2008) has focused on such joint cross-layer optimization. They also show that in order to achieve stability, back-pressure algorithms should exploit all possible paths between the source and destination. This ensures perfect load balancing and bounded queues when further assisted using the appropriate rate control and scheduling modules.

One of the reasons for which routing is assumed to preset in back-pressure algorithms is that exploiting all paths has multiple challenges. First, the message overhead require in order to determine these paths in distributed fashion prohibitively large. Second, since the packets are being forwarded on longer paths, the end-to-end delivery suffers larger delays. While multiple paths are necessary for queue stability under high loads, Ying et al. (2009) showed that they results into unnecessary reduction of delay when the load is moderate to low. This was shown by Ying et al. (2009) using a small Manhattan Street Network as an example, in which nodes generate data with probability of λ at the beginning of every time slot. The packets are forwarded to a randomly chosen destination.

They observe, surprisingly, that with small values of λ (0.1 and less), the delay is very high, and falls rapidly to a minimum at around $\lambda = 0.1$, rising slowly with subsequent increase in λ. This behavior is due to the fact that at lower traffic load, exploiting all possible paths for forwarding data becomes unnecessary and in fact disadvantageous in terms of delay. To circumvent the problem, Ying et al. (2009) propose algorithms in which delay is very low with low values of λ, with the expected slow rise, now monotonically, with the traffic load without compromising the throughput optimality. To do this, it proposes two optimization problems:(i) packets of each flow in must not traverse a fixed number of hops while being delivered using back-pressure routing, (ii) packets of each flow should be delivered in minimum number of total hops. Both the problems are based on the central idea that increase in number of hops results into increased delay. The solutions are designed in which the algorithms make use of the shortest path between source-destination in order to minimize the number of hops of back-pressure routes. Note that in stead of maintaining PDQs at each node, the schemes proposes to separate each PDQ on the bases of their hop-constraints. These added queue management is shown to be sufficient for route management while achieving throughput optimality.

To further understand the delay performance of back-pressure scheme, Bui et al. (2009) identify an additional issue that contributes to the poor delay performance of back-pressure scheme. It states that the delay can be high also due to the fact that each node has to maintain a separate queue for each flow passing through it. It shows that when routing is fixed, shadow queues Bui et al. (2008) can be used to reduce the queue management complexity. When using shadow queues, it is shown that maintaining per-neighbor queues are sufficient instead of maintaining per-flow queues. The proposed solution also explores shorter paths first in order to achieve a desirable delay performance. If only the throughput optimality can not be met using the initial set of paths, the algorithm explores more paths.

The throughput optimality of back-pressure scheme is based on an assumption that traffic flows are stationary ergodic processes. In a more realistic scenarios where traffic flows start and end many times during the operation of the mesh network, the stability of the queues is not properly understood. This was first explored by van de Ven et al. (2009) which showed that when traffic processes are not stationary and ergodic, back-pressure framework can not guarantee queue stability and throughput optimality. The problem was further identified by Ji et al. (2011). It defines the following problems:

The Last Packet Problem: Consider a flow for which the source stops generating packets but the last generated packet is on the way towards the destination. The queue at a link holding the packet does not get scheduled. This is because queues of other flows that are already generating packets at a faster rate are always preferred in scheduling, and the queue with the last packet starves due to insignificant backlog weight.

To mitigate the last packet problem, Andrews et al. (2001, 2004), Mekkittikul and McKeown (1996), Sadiq and De Veciana (2009) and Shakkottai and Stolyar (2002) consider utilizing Head-of-line packet delay as the weight for scheduling instead of the back-log. Though this intuitive solution can solve the last packet problem, the throughput optimality and queue stability is not proven. Ji et al. (2011) use difference of sojourn times between Head-of-line packets instead of difference between queue lengths of queues for the purpose of link scheduling. It is shown that this not only solves the last packet problem but also does so while maintaining queue stability and throughput optimality. Also, the presented algorithm has shown to have superior delay performance due to the fact that packet sojourn times are already considered in the scheduling process.

A real-world implementation of back-pressure framework is not widely adopted because of two reasons: (1) a large number of message passing is required in order to correctly implement it in distributed manner, (2) its direct implementation requires modifying TCP and MAC modules of current protocol stack which is always less desirable. Radunović et al. (2008) implement a simple yet efficient variant of back-pressure framework such that message overhead is smaller while utility maximization goals are achieved. Similarly, Moeller et al. (2010) demonstrated that replacing FIFO queues of back-pressure framework with LIFO queues can improve the delay performance by upto 75 %. This is further used for designing a back-pressure based data collection protocol for wireless sensor networks. Along the same lines, Ryu et al. (2010a,b), Ying et al. (2008) use back-pressure framework for intermittently connected networks and Internet-like communication network.

Towards real-world implementation of back-pressure in 802.11 type of MAC protocols, Laufer et al. (2011) presented the first solution named XPRESS. It implements

Fig. 5.7 Delay characteristic of XPRES, a real-world attempt at back-pressure

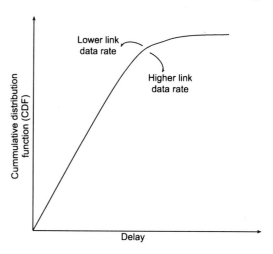

a complete back-pressure framework using TDMA based MAC which utilizes customized data and control planes. In a single flow multi-hop routing path experiment, Laufer et al. (2011) showed that XPRESS can in fact utilize all possible loop-free paths between the source and the destination. Their results show that the resultant throughput increase of XPRESS compared to 802.11 can be as high as 128 %. The delay performance is observed to have the general nature shown in Fig. 5.7. Higher delays and longer routing paths are observed when the input rate is low because back-pressure framework explores a large number of paths while forwarding the data. With real-world hardware able to achieve such significant increase in performance, back-pressure strategy indeed holds a great promise for wireless mesh network in future.

5.11 Cognitive Mesh Networks

As we saw in Chap. 2, cognitive mesh network design requires tight coupling of sensing module and upper layers which makes the design of upper layers cross-layer. Here, we discuss routing and congestion control problems in cognitive mesh networks.

5.11.1 Routing

In cognitive mesh network, mesh nodes act as secondary stations that operate alongside with primary network. The primary network can be any other network such as cellular network, TV broadcast network etc.

The routing problem is especially challenging in cognitive mesh networks because the availability of links depend on activity of primary users. Even if there exist a way to correctly sense primary users' activity, availability of links change quickly which requires that routing paths are also altered accordingly. The stability of routes is an important design objective to be considered while designing routing protocols for cognitive mesh networks.

There are many different routing protocols proposed for cognitive mesh network. Khalife et al. (2008) propose a source routing protocol that attempts to compute the most probable path between source and destination. It proposes a link metric that considers the interference of primary user and uses it to derive link availability. Also, the path found by the routing path is expected to achieve a certain pre-established capacity. It is also important to consider that primary nodes might be mobile which alters their spatial characteristics as well. To tackle the issue Beltagy et al. (2011) propose a multipath routing in which derived routing paths are expected to more and more divergent so that primary user activity might not affect multiple paths at the same time.

It is necessary to study the underlying connectivity of secondary network in order to derive a routing protocol. Ren et al. (2009a,b, 2011) studied the connectivity using continuum percolation, and provided expression to capture the relation between number of primary neighbors in surrounding and temporal opportunities of transmission for secondary users. Abbagnale and Cuomo (2010a,b) use a similar approach for deriving secondary user connectivity that is based on algebraic connectivity. The local notion of such connectivity reduces to a weight which shows how much a link is likely to be affected by nearby primary user's activity. This is used to derive a distributed routing protocol that can computer robust and high throughput routing paths.

5.11.2 Congestion Control

The primary reason why congestion control and end-to-end reliability is important concern in cognitive mesh network is because packets can face long round trip time due to inability of intermediate secondary nodes to forward the packets immediately. This can happen when a secondary node's transmission is preempted by primary node transmission. This requires the secondary node to buffer the packet which in turn affects the end-to-end latency. This requires that an end-to-end congestion control protocol coordinates with spectrum sensing module. Chowdhury et al. (2009) propose a transport protocol to address these issues for cognitive mesh network. In the protocol, rate control module utilizes information from the sensing module in order to control the input rate by closely monitoring spectrum availability. The protocol is based on explicit feedback provided by intermediate nodes to the source node for congestion and rate control. Kim et al. (2011) analyze the TCP throughput of a cognitive mesh that acts as a secondary network to already deployed primary network of WiFi based access points.

The problems of spectrum sensing, channel selection, routing, medium access, congestion control are strongly correlated in cognitive mesh network. The being true, only some research has been devoted to design of such cross-layer protocols, and the joint design in cognitive mesh network largely remains unexplored open research direction.

References

Abbagnale A, Cuomo F (2010a) Connectivity-driven routing for cognitive radio ad-hoc networks. In: 7th Annual IEEE communications society conference on sensor mesh and ad hoc communications and networks (SECON), pp 1–9, doi:10.1109/SECON.2010.5508269

Abbagnale A, Cuomo F (2010b) Gymkhana: a connectivity-based routing scheme for cognitive radio ad hoc networks. In: INFOCOM IEEE conference on computer communications workshops, pp 1–5. doi:10.1109/INFOCOMW.2010.5466618

Akyol U, Andrews M, Gupta P, Hobby J, Saniee I, Stolyar A (2008) Joint scheduling and congestion control in mobile ad-hoc networks. In: INFOCOM 2008. The 27th conference on computer communications. IEEE, pp 619–627. doi:10.1109/INFOCOM.2008.111

Alicherry M, Bhatia R, Li LE (2005) Joint channel assignment and routing for throughput optimization in multi-radio wireless mesh networks. In: MobiCom'05: proceedings of the 11th annual international conference on mobile computing and networking. ACM, New York, pp 58–72. doi:10.1145/1080829.1080836

Andrews M, Kumaran K, Ramanan K, Stolyar A, Whiting P, Vijayakumar R (2001) Providing quality of service over a shared wireless link. Commun Mag IEEE 39(2):150–154. doi:10.1109/35.900644

Andrews M, Kumaran K, Ramanan K, Stolyar A, Vijayakumar R, Whiting P (2004) Scheduling in a queuing system with asynchronously varying service rates. Probab Eng Inf Sci 18:191–217. doi:10.1017/S0269964804182041, http://dl.acm.org/citation.cfm?id=985908.985912

Badawy G, Sayegh A, Todd T (2008a) Energy aware provisioning in solar powered wlan mesh networks. In: Proceedings of 17th international conference on computer communications and networks, ICCCN'08, pp 1–6. doi:10.1109/ICCCN.2008.ECP.96

Badawy G, Sayegh A, Todd T (2008b) Solar powered wlan mesh network provisioning for temporary deployments. In: Wireless communications and networking conference, 2008. WCNC 2008. IEEE, pp 2271–2276. doi:10.1109/WCNC.2008.401

Badawy G, Sayegh A, Todd T (2009a) Fair flow control in solar powered wlan mesh networks. In: Wireless communications and networking conference. WCNC 2009, IEEE, pp 1–6. doi:10.1109/WCNC.2009.4917720

Badawy G, Sayegh A, Todd T (2009b) Managing traffic growth in solar powered wireless mesh networks. In: global telecommunications conference. GLOBECOM 2009. IEEE, pp 1–6. doi:10.1109/GLOCOM.2009.5425730

Badawy G, Sayegh A, Todd T (2010) Energy provisioning in solar-powered wireless mesh networks. IEEE Trans Veh Technol 59(8):3859–3871. doi:10.1109/TVT.2010.2064797

Beltagy I, Youssef M, El-Derini M (2011) A new routing metric and protocol for multipath routing in cognitive networks. In: Wireless communications and networking conference (WCNC), 2011. IEEE, pp 974–979. doi:10.1109/WCNC.2011.5779268

Bhatia R, Kodialam M (2004) On power efficient communication over multi-hop wireless networks: joint routing, scheduling and power control. In: INFOCOM 2004. twenty-third annual joint conference IEEE computer communications societies, vol 2, pp 1457–1466

Bhatia R, Li L (2007) Throughput optimization of wireless mesh networks with mimo links. In: INFOCOM 2007, 26th IEEE international conference on computer communications. pp 2326–2330. doi:10.1109/INFOCOM.2007.274

Brzezinski A, Zussman G, Modiano E (2006) Enabling distributed throughput maximization in wireless mesh networks: a partitioning approach. In: MobiCom'06: proceedings of the 12th annual international conference on mobile computing and networking, ACM, New York, pp 26–37. doi:10.1145/1161089.1161094

Bui BYL, Srikant R, Stolyar A (2008) Optimal resource allocation for multicast sessions in multihop wireless networks. J Comput Biol 19(6):785–795

Bui L, Srikant R, Stolyar A (2009) Novel architectures and algorithms for delay reduction in back-pressure scheduling and routing. In: INFOCOM 2009, IEEE, pp 2936–2940. doi:10.1109/INFOCOM.2009.5062262

Capone A, Carello G (2006) Scheduling optimization in wireless mesh networks with power control and rate adaptation. In: 3rd Annual IEEE communications society on sensor and ad hoc communications and networks, SECON'06 2006, vol 1, pp 138–147. doi:10.1109/SAHCN.2006.288418

Chen CC, Lee DS (2006) A joint design of distributed QoS scheduling and power control for wireless networks. In: INFOCOM 2006, 25th IEEE international conference on computer communications proceedings, pp 1–12. doi:10.1109/INFOCOM.2006.294

Chen L, Low SH, Chiang M, Doyle JC (2006) Cross-layer congestion control, routing and scheduling design in ad hoc wireless networks. In: Proceedings of INFOCOM 2006. 25th IEEE international conference on computer communications, pp 1–13. doi:10.1109/INFOCOM.2006.142

Chiu DM, Jain R (1989) Analysis of the increase and decrease algorithms for congestion avoidance in computer networks. Comput Netw ISDN Syst 17(1):1–14. doi:10.1016/0169-7552(89)90019-6

Chowdhury K, Di Felice M, Akyildiz I (2009) Tp-crahn: A transport protocol for cognitive radio ad-hoc networks. In: Proceedings of INFOCOM 2009, IEEE, pp 2482–2490. doi:10.1109/INFOCOM.2009.5062176

Cruz R, Santhanam A (2003) Optimal routing, link scheduling and power control in multihop wireless networks. In: INFOCOM 2003, twenty-second annual joint conference of the IEEE computer and communications societies IEEE, vol 1, pp 702–711. doi:10.1109/INFOCOM.2003.1208720

Deuskar G, Pathak P, Dutta R, (2012) Packet aggregation based back-pressure scheduling in multi-hop wireless networks. In: IEEE 2012, wireless communications and networking conference: MAC and cross-layer design (IEEE WCNC 2012 Track 2 MAC), Paris

Durvy M, Thiran P (2006) A packing approach to compare slotted and non-slotted medium access control. In: Proceedings of the INFOCOM 2006, 25th IEEE international conference on computer communications, pp 1–12. doi:10.1109/INFOCOM.2006.251

ElBatt T, Ephremides A (2004) Joint scheduling and power control for wireless ad hoc networks. Wirel Commun IEEE Trans 3(1):74–85. doi:10.1109/TWC.2003.819032

Eryilmaz A, Srikant R (2005) Fair resource allocation in wireless networks using queue-length-based scheduling and congestion control. In: Proceedings of the INFOCOM 2005, 24th annual joint conference of the IEEE computer and communications societies. IEEE, vol 3, pp 1794–1803. doi:10.1109/INFOCOM.2005.1498459

Eryilmaz A, Srikant R (2006) Joint congestion control, routing, and mac for stability and fairness in wireless networks. IEEE J Sel Areas Commun, 24(8):1514–1524. doi:10.1109/JSAC.2006.879361

Fan KW, Zheng Z, Sinha P (2008) Steady and fair rate allocation for rechargeable sensors in perpetual sensor networks. In: Proceedings of the 6th ACM conference on embedded network sensor systems, ACM, New York, SenSys'08, pp 239–252. doi:10.1145/1460412.1460436,10.1145/1460412.1460436

Farbod A, Todd T (2007) Resource allocation and outage control for solar-powered wlan mesh networks. IEEE Trans Mobile Comput 6(8):960–970. doi:10.1109/TMC.2007.1079

Foukalas F, Gazis V, Alonistioti N (2008) Cross-layer design proposals for wireless mobile networks: a survey and taxonomy. IEEE Commun Surv Tutor 10(1):70–85. doi:10.1109/COMST.2008.4483671

Ganguly S, Navda V, Kim K, Kashyap A, Niculescu D, Izmailov R, Hong S, Das S (2006) Performance optimizations for deploying voip services in mesh networks. IEEE J Sel Areas Commun 24(11):2147–2158. doi:10.1109/JSAC.2006.881594

Georgiadis L, Neely MJ, Tassiulas L (2006) Resource allocation and cross-layer control in wireless networks. Found Trends Netw 1:1–144. doi:10.1561/1300000001, http://dl.acm.org/citation.cfm?id=1166401.1166402

Goussevskaia O, Oswald YA, Wattenhofer R (2007) Complexity in geometric SINR. In: Mobi-Hoc'07: proceedings of the 8th ACM international symposium on mobile ad hoc networking and computing, ACM, New York, pp 100–109. doi:10.1145/1288107.1288122

Huang S, Dutta R (2006) Design of wireless mesh networks under the additive interference model. In: Proceedings of the 15th international conference on computer communications and networks, ICCCN 2006, pp 253–260. doi:10.1109/ICCCN.2006.286282

Jain A, Gruteser M, Neufeld M, Grunwald D (2003) Benefits of packet aggregation in ad-hoc wireless network. Technical report, University of Colorado at Boulder

Ji B, Joo C, Shroff N (2011) Delay-based back-pressure scheduling in multi-hop wireless networks. In: Proceedings of IEEE INFOCOM, pp 2579–2587. doi:10.1109/INFCOM.2011.5935084

Jiang L, Walrand J (2010) A distributed CSMA algorithm for throughput and utility maximization in wireless networks. IEEE/ACM Trans Netw 18(3):960–972. doi:10.1109/TNET.2009.2035046

Jiang LB, Liew SC (2008) Improving throughput and fairness by reducing exposed and hidden nodes in 802.11 networks. IEEE Trans Mobile Comput. 7(1):34–49. doi:10.1109/TMC.2007.1070

Kansal A, Potter D, Srivastava MB (2004) Performance aware tasking for environmentally powered sensor networks. In: Proceedings of the joint international conference on measurement and modeling of computer systems, ACM, New York, SIGMETRICS'04/Performance '04, pp 223–234. http://doi.acm.org/10.1145/1005686.1005714,http://doi.acm.org/10.1145/1005686.1005714

Kanth K, Ansari S, Melikri M (2002) Performance enhancement of tcp on multihop ad hoc wireless networks. In: IEEE international conference on personal wireless communications, pp 90–94. doi:10.1109/ICPWC.2002.1177252

Kashyap A, Lee K, Kalantari M, Khuller S, Shayman M (2007a) Integrated topology control and routing in wireless optical mesh networks. Comput Netw 51(15):4237–4251. http://dx.doi.org/10.1016/j.comnet.2007.05.006

Kashyap A, Sengupta S, Bhatia R, Kodialam M (2007b) Two-phase routing, scheduling and power control for wireless mesh networks with variable traffic. In: SIGMETRICS'07: proceedings of the 2007 ACM SIGMETRICS international conference on measurement and modeling of computer systems, ACM, New York, pp 85–96. http://doi.acm.org/10.1145/1254882.1254893

Kelly F, Maulloo A, Tan D (1998) Rate control in communication networks: shadow prices, proportional fairness and stability. J oper res soc, vol. 49:237–252. http://citeseer.ist.psu.edu/kelly98rate.html

Khalife H, Ahuja S, Malouch N, Krunz M (2008) Probabilistic path selection in opportunistic cognitive radio networks. In: Global telecommunications conference, IEEE GLOBECOM 2008. IEEE, pp 1–5. doi:10.1109/GLOCOM.2008.ECP.931

Kim W, Kassler A, Gerla M (2011) Tcp performance in cognitive multi-radio mesh networks. In: Proceedings of the 4th international conference on cognitive radio and advanced spectrum management ACM, New York, CogART'11, pp 44:1–6. doi:10.1145/2093256.2093300, http://doi.acm.org/10.1145/2093256.2093300

Kliazovich D, Granelli F (2008) Packet concatenation at the ip level for performance enhancement in wireless local area networks. Wirel Netw 14(4):519–529. http://dx.doi.org/10.1007/s11276-006-0734-6

Kodialam M, Nandagopal T (2003) Characterizing the achievable rates in multi-hop wireless networks: the joint routing and scheduling problem. MobiCom'03: proceedings of the 9th annual international conference on mobile computing and networking

Kodialam M, Nandagopal T (2005) Characterizing the capacity region in multi-radio multi-channel wireless mesh networks. MobiCom'05: proceedings of the 11th annual international conference on mobile computing and networking

Kulkarni G, Raghunathan V, Srivastava M (2004) Joint end-to-end scheduling, power control and rate control in multi-hop wireless networks. Global Telecommunications Conference, GLOBE-COM'04, IEEE. vol 5, pp 3357–3362. doi:10.1109/GLOCOM.2004.1378971

Laufer R, Salonidis T, Lundgren H, Le Guyadec P (2011) Xpress: a cross-layer backpressure architecture for wireless multi-hop networks. In: Proceedings of the 17th annual international conference on mobile computing and networking, ACM, New York, MobiCom'11, pp 49–60. doi:10.1145/2030613.2030620, 10.1145/2030613.2030620

Lee J, Lee J, Lee K, Chong S (2010) Commoncode: a code reuse platform for co-simulation and experimentation. In: Proceedings of the ACM CoNEXT Student Workshop, ACM, New York, CoNEXT'10 Student Workshop, pp 13:1–13:2, http://doi.acm.org/10.1145/1921206.1921221, http://doi.acm.org/10.1145/1921206.1921221

Li Y, Ephremides A (2007) A joint scheduling, power control, and routing algorithm for ad hoc wireless networks. Ad Hoc Netw 5(7):959–973. http://dx.doi.org/10.1016/j.adhoc.2006.04.005

Lim H, Lim C, Hou JC (2006) A coordinate-based approach for exploiting temporal-spatial diversity in wireless mesh networks. In: MobiCom'06: proceedings of the 12th annual international conference on mobile computing and networking, ACM, New York, pp 14–25. http://doi.acm.org/10.1145/1161089.1161093

Lin X, Shroff N (2004) Joint rate control and scheduling in multihop wireless networks. In: 43rd IEEE conference on decision and control, CDC, vol. 2, pp 1484–1489. doi:10.1109/CDC.2004.1430253

Lin X, Shroff N, Srikant R (2006) A tutorial on cross-layer optimization in wireless networks. IEEE J Sel Areas Commun 24(8):1452–1463. doi:10.1109/JSAC.2006.879351

Liu H, Yu H, Liu X, Chuah CN, Mohapatra P (2007) Scheduling multiple partially overlapped channels in wireless mesh networks. In: ICC'07 IEEE international conference on communications, pp 3817–3822. doi:10.1109/ICC.2007.629

Liu J, Yi Y, Proutiere A, Chiang M, Poor HV (2010) Towards utility-optimal random access without message passing. Wirel Commun Mob Comput 10:115–128. doi:10.1002/wcm.v10:1, http://dl.acm.org/citation.cfm?id=1687265.1687268

Marbach P, Eryilmaz A (2008) A backlog-based CSMA mechanism to achieve fairness and throughput-optimality in multihop wireless networks. In: 46th annual Allerton conference on Communication, Control, and Computing, pp 768–775. doi:10.1109/ALLERTON.2008.4797635

Mekkittikul A, McKeown N (1996) A starvation-free algorithm for achieving 100% throughput in an input- queued switch. In: Proceedings of the ICCCN 96, Stanford University, CA, Oct 1996

Mishra A, Shrivastava V, Banerjee S, Arbaugh W (2006) Partially overlapped channels not considered harmful. SIGMETRICS Perform Eval Rev 34(1):63–74. http://doi.acm.org/10.1145/1140103.1140286

Moeller S, Sridharan A, Krishnamachari B, Gnawali O (2010) Routing without routes: the backpressure collection protocol. In: Proceedings of the 9th ACM/IEEE international conference on information processing in sensor networks IPSN '10. ACM, New York, pp 279–290. http://doi.acm.org/10.1145/1791212.1791246, http://doi.acm.org/10.1145/1791212.1791246

Mohsenian Rad A, Wong V (2007) Partially overlapped channel assignment for multi-channel wireless mesh networks. In: IEEE International conference on communications, ICC'07, pp 3770–3775. doi:10.1109/ICC.2007.621

Moscibroda T, Wattenhofer R (2006) The complexity of connectivity in wireless networks. In: INFOCOM 2006, 25th IEEE international conference on computer communications proceedings, pp 1–13. doi:10.1109/INFOCOM.2006.23

Moscibroda T, Wattenhofer R, Zollinger A (2006) Topology control meets SINR: the scheduling complexity of arbitrary topologies. In: MobiHoc '06: proceedings of the 7th ACM international symposium on Mobile ad hoc networking and computing, ACM, New York, pp 310–321. http://doi.acm.org/10.1145/1132905.1132939

Moscibroda T, Oswald Y, Wattenhofer R (2007) How optimal are wireless scheduling protocols? In: INFOCOM 2007, 26th IEEE International conference on computer communications, IEEE, pp 1433–1441. doi:10.1109/INFCOM.2007.169

Moser C, Brunelli D, Thiele L, Benini L (2007) Real-time scheduling for energy harvesting sensor nodes. Real-Time Syst 37:233–260. doi:10.1007/s11241-007-9027-0, http://dl.acm.org/citation.cfm?id=1295875.1295893

Na C, Chen J, Rappaport T (2006) Measured traffic statistics and throughput of ieee 802.11b public wlan hotspots with three different applications. IEEE Trans Wirel Commun 5(11):3296–3305. doi:10.1109/TWC.2006.05043

Nardelli B, Lee J, Lee K, Yi Y, Chong S, Knightly E, Chiang M (2011) Experimental evaluation of optimal CSMA. In: Proceedings of IEEE INFOCOM, pp 1188–1196. doi:10.1109/INFCOM.2011.5934897

Neely M (2006) Energy optimal control for time-varying wireless networks. IEEE Trans Inform Theory 52(7):2915–2934. doi:10.1109/TIT.2006.876219

Neely M, Modiano E, Li CP (2005a) Fairness and optimal stochastic control for heterogeneous networks. In: Proceedings of the IEEE 24th annual joint conference of the IEEE computer and communications societies, INFOCOM 2005, vol. 3, pp 1723–1734. doi:10.1109/INFCOM.2005.1498453

Neely M, Modiano E, Rohrs C (2005b) Dynamic power allocation and routing for time-varying wireless networks. IEEE J Sel Areas Commun 23(1):89–103. doi:10.1109/JSAC.2004.837349(410)23

Neely MJ, Urgaonkar R (2009) Optimal backpressure routing for wireless networks with multi-receiver diversity. Ad Hoc Netw 7:862–881. doi:10.1016/j.adhoc.2008.07.009, http://dl.acm.org/citation.cfm?id=1508337.1508838

Ni J, Tan B, Srikant R (2012) Q-csma: Queue-length-based csma/ca algorithms for achieving maximum throughput and low delay in wireless networks. IEEE/ACM Trans Netw 99:1. doi:10.1109/TNET.2011.2177101

Niyato D, Hossain E, Fallahi A (2007) Sleep and wakeup strategies in solar-powered wireless sensor/mesh networks: performance analysis and optimization. IEEE Trans Mobile Comput 6(2):221–236. doi:10.1109/TMC.2007.30

Noh DK, Wang L, Yang Y, Le HK, Abdelzaher T (2009) Minimum variance energy allocation for a solar-powered sensor system. In: Proceedings of the 5th IEEE international conference on distributed computing in sensor systems DCOSS'09, Springer, Berlin, pp 44–57, http://dx.doi.org/10.1007/978-3-642-02085-8_4, http://dx.doi.org/10.1007/978-3-642-02085-8_4

Pathak P, Dutta R (2010) A survey of network design problems and joint design approaches in wireless mesh networks. IEEE Commun Surv Tutor PP(99):1–33. doi:10.1109/SURV.2011.060710.00062

Pathak PH, Gupta D, Dutta R (2008) Loner links aware routing and scheduling wireless mesh networks. In: ANTS 2008 2nd advanced networking and telecommunications system conference, IEEE

Radunović B, Gkantsidis C, Gunawardena D, Key P (2008) Horizon: balancing tcp over multiple paths in wireless mesh network. In: Proceedings of the 14th ACM international conference on mobile computing and networking MobiCom'08, ACM, New York, pp 247–258 http://doi.acm.org/10.1145/1409944.1409973, http://doi.acm.org/10.1145/1409944.1409973

Raghavendra R, Jardosh A, Belding E, Zheng H (2006) Ipac: Ip-based adaptive packet concatenation for multihop wireless networks. In: Fortieth Asilomar conference on signals, systems and computers, ACSSC '06, pp 2147–2153. doi:10.1109/ACSSC.2006.355148

Rajagopalan S, Shah D (2008) Distributed algorithm and reversible network. In: 42nd annual conference on information sciences and systems. CISS 2008, pp 498–502, doi:10.1109/CISS.2008.4558577

Rajagopalan S, Shah D, Shin J (2009) Network adiabatic theorem: an efficient randomized protocol for contention resolution. In: Proceedings of the eleventh international joint conference on measurement and modeling of computer systems, ACM, New York, SIGMETRICS'09, pp 133–144. http://doi.acm.org/10.1145/1555349.1555365, http://doi.acm.org/10.1145/1555349.1555365

Raniwala A, cker Chiueh T (2005) Architecture and algorithms for an ieee 802.11-based multi-channel wireless mesh network. In:proceedings of IEEE 24th annual joint conference of the IEEE computer and communications societies INFOCOM 2005, vol. 3, pp 2223–2234. doi:10.1109/INFCOM.2005.1498497

Raniwala A, Gopalan K, Chiueh T (2004) Centralized channel assignment and routing algorithms for multi-channel wireless mesh networks. SIGMOBILE Mob Comput Commun Rev 8(2):50–65. http://doi.acm.org/10.1145/997122.997130

Ren W, Zhao Q, Swami A (2009a) Connectivity of cognitive radio networks: proximity vs. opportunity. In: Proceedings of the 2009 ACM workshop on cognitive radio networks CoRoNet'09, ACM, New York, pp 37–42. doi:10.1145/1614235.1614245,10.1145/1614235.1614245

Ren W, Zhao Q, Swami A (2009b) Power control in cognitive radio networks: how to cross a multi-lane highway. IEEE J Sel Areas Commun 27(7):1283–1296. doi:10.1109/JSAC.2009.090923

Ren W, Zhao Q, Swami A (2011) On the connectivity and multihop delay of ad hoc cognitive radio networks. IEEE J Sel Areas Commun 29(4):805–818. doi:10.1109/JSAC.2011.110412

Riggio R, Miorandi D, De Pellegrini F, Granelli F, Chlamtac I (2008) A traffic aggregation and differentiation scheme for enhanced qos in ieee 802.11-based wireless mesh networks. Comput Commun 31(7):1290–1300. http://dx.doi.org/10.1016/j.comcom.2008.01.037

Ryu J, Bhargava V, Paine N, Shakkottai S (2010a) Back-pressure routing and rate control for icns. In: Proceedings of the sixteenth annual international conference on mobile computing and networking MobiCom'10, ACM, New York, pp 365–376, http://doi.acm.org/10.1145/1859995.1860037,http://doi.acm.org/10.1145/1859995.1860037

Ryu J, Ying L, Shakkottai S (2010b) Back-pressure routing for intermittently connected networks. In: INFOCOM, 2010 proceedings IEEE, pp 1–5. doi:10.1109/INFCOM.2010.5462224

Sadiq B, De Veciana G (2009) Throughput optimality of delay-driven maxweight scheduler for a wireless system with flow dynamics. In: Proceedings of the 47th annual Allerton conference on communication, control, and computing, IEEE Press, Piscataway, NJ, Allerton'09, pp 1097–1102. http://dl.acm.org/citation.cfm?id=1793974.1794164

Sayegh A, Todd T (2007) Energy management in solar powered wlan mesh nodes using online meteorological data. In: IEEE international conference on communications, 2007. ICC'07, pp 3811–3816. doi:10.1109/ICC.2007.628

Sayegh A, Ghosh S, Todd T (2008) Optimal node placement in hybrid solar powered wlan mesh networks. In: Wireless communications and networking conference, 2008. WCNC 2008. IEEE, pp 2277–2282. doi:10.1109/WCNC.2008.402

Shakkottai S, Stolyar A (2002) Scheduling for multiple flows sharing a time-varying channel: the exponential rule. In: Suhov Yu.M (ed) Analytic methods in applied probability. American Mathematical Society, Providence (2002)

Shariat M, Quddus A, Ghorashi S, Tafazolli R (2009) Scheduling as an important cross-layer operation for emerging broadband wireless systems. Commun Surv Tutor IEEE 11(2):74–86. doi:10.1109/SURV.2009.090206

Sharma N, Gummeson J, Irwin D, Shenoy P (2010) Cloudy computing: leveraging weather forecasts in energy harvesting sensor systems. In: 7th annual IEEE communications society conference on sensor mesh and ad hoc communications and networks (SECON), pp 1–9. doi:10.1109/SECON.2010.5508260

Sharma N, Sharma P, Irwin D, Shenoy P (2011) Predicting solar generation from weather forecasts using machine learning. In: IEEE international conference on smart grid communications (SmartGridComm), pp 528–533. doi:10.1109/SmartGridComm.2011.6102379

Stolyar AL (2005) Maximizing queueing network utility subject to stability: Greedy primal-dual algorithm. Queueing Syst Theory Appl 50:401–457. doi:10.1007/s11134-005-1450-0, http://dl.acm.org/citation.cfm?id=1085031.1085090

Tang D, Baker M (2000) Analysis of a local-area wireless network. In: MobiCom'00: proceedings of the 6th annual international conference on mobile computing and networking, ACM, New York, pp 1–10. http://doi.acm.org/10.1145/345910.345912

Tang J, Xue G, Zhang W (2005) Interference-aware topology control and QoS routing in multi-channel wireless mesh networks. In: MobiHoc'05: proceedings of the 6th ACM international symposium on mobile ad hoc networking and computing, ACM, New York, pp 68–77. http://doi. acm.org/10.1145/1062689.1062700

Tarn WH, Tseng YC (2007) Joint multi-channel link layer and multi-path routing design for wireless mesh networks. In: INFOCOM 2007 26th IEEE international conference on computer communications IEEE, pp 2081–2089. doi:10.1109/INFCOM.2007.241

Tassiulas L, Ephremides A (1990) Stability properties of constrained queueing systems and scheduling policies for maximum throughput in multihop radio networks. In: Proceedings of the 29th IEEE conference on decision and control, vol. 4, pp 2130–2132. doi:10.1109/CDC.1990.204000

Tassiulas L, Ephremides A (1992) Jointly optimal routing and scheduling in packet ratio networks. IEEE Trans Inform Theory 38(1):165–168. doi:10.1109/18.108264

Vargas E, Sayegh A, Todd T (2007) Shared infrastructure power saving for solar powered ieee 802.11 wlan mesh networks. In: IEEE international conference on communications. ICC'07, pp 3835–3840. doi:10.1109/ICC.2007.632

van de Ven P, Borst S, Shneer S (2009) Instability of maxweight scheduling algorithms. In: INFOCOM 2009, IEEE, pp 1701–1709. doi:10.1109/INFCOM.2009.5062089

Wang W, Liu X, Krishnaswamy D (2007) Robust routing and scheduling in wireless mesh networks. In: 4th Annual IEEE communications society conference on sensor, mesh and ad hoc communications and networks, SECON'07, pp 471–480. doi:10.1109/SAHCN.2007.4292859

Warrier A, Janakiraman S, Ha S, Rhee I (2009) Diffq: practical differential backlog congestion control for wireless networks. In: INFOCOM 2009, IEEE, pp 262–270. doi:10.1109/INFCOM. 2009.5061929

Yeh E, Berry R (2007a) Throughput optimal control of cooperative relay networks. IEEE Trans Inform Theory 53(10):3827–3833, doi:10.1109/TIT.2007.904978

Yeh E, Berry R (2007b) Throughput optimal control of wireless networks with two-hop cooperative relaying. In: IEEE international symposium on information theory, ISIT 2007, pp 351–355. doi:10.1109/ISIT.2007.4557090

Ying L, Srikant R, Towsley D (2008) Cluster-based back-pressure routing algorithm. In: INFOCOM 2008. The 27th conference on computer communications. IEEE, pp 484–492. doi:10. 1109/INFOCOM.2008.96

Ying L, Shakkottai S, Reddy A (2009) On combining shortest-path and back-pressure routing over multihop wireless networks. In: INFOCOM 2009, IEEE, pp 1674–1682. doi:10.1109/INFCOM. 2009.5062086

Ying L, Shakkottai S, Reddy A, Liu S (2010) On combining shortest-path and back-pressure routing over multihop wireless networks. IEEE/ACM Trans Netw, PP(99):1. doi:10.1109/TNET.2010. 2094204

Z Wu DR S Ganu (2006) Irma: integrated routing and MAC scheduling in multi-hop wireless mesh networks. In: WiMesh 2006, 2nd IEEE workshop on wireless mesh networks

Zefreh M, Badawy G, Todd T (2010) Position aware node provisioning for solar powered wireless mesh networks. In: Global telecommunications conference (GLOBECOM 2010), 2010 IEEE, pp 1–6. doi:10.1109/GLOCOM.2010.5683196

Zhai H, Fang Y (2005) A distributed packet concatenation scheme for sensor and ad hoc networks. In: Military communications conference, 2005. MILCOM 2005. IEEE, vol. 3, pp 1443–1449. doi:10.1109/MILCOM.2005.1605880

Zhang J, Wu H, Zhang Q, Li B (2005) Joint routing and scheduling in multi-radio multi-channel multi-hop wireless networks. In: 2nd international conference on broadband networks, vol. 1, pp 631–640. doi:10.1109/ICBN.2005.1589668

Chapter 6
Survivability of Mesh Networks

6.1 A Primer on Network Survivability

The fundamental functionality of any networked system is to enable communication among the desired entities. This functionality is perceived by network users as a service, and any disruption in the service (partial or complete) implies failure of the network as a system. In last four decades, disruption of networking service has been studied in various forms such as reliability, dependability, fault-tolerance, survivability, trustworthiness, robustness etc. The term *reliability* was introduced in the military context in order to define service disruption of a stand-alone computing system due to hardware failures. Along the same lines, *fault-tolerance* was introduced as a concept to capture the effects of software failures (mostly due to design issues). Later, as the systems became more and integrated, the concept of *dependability* (often used as an umbrella term) evolved to encompass hardware and software failures and its effects. With the emergence of the Internet, the need to study failures in a networked computing system became apparent, which gave birth to another (rather general) term, *survivability*. In the last decade, failures due to malicious attacks and intentional disruptions have given rise to the concepts of *security* and *trustworthiness*. Other definitions of service expectations (e.g. Quality of Service of Quality of Experience) have evolved in parallel but their meanings have remained less standard, and usually more complex (and emerging from multiple simpler quantities).

Understanding service disruption is complex especially in networks because of the large space of inter-dependencies and levels of error propagation. As a first step, a fault typically refers to a malfunctioning of an atomic component. Here, malfunctioning means that the functionality of the component deviates from its expected or specified behavior. Until a fault alters the system's expected behavior, it is denoted dormant. Due to changed circumstances (external or internal to the network), the fault can become active and can cause an error. An error is a state of the network which may or may not lead to a service disruption. Apart from this, an error can propagate to other components due to functional dependencies. When the propagation reaches a degree that affects the level of a perceived network service, it is typically referred

P. H. Pathak and R. Dutta, *Designing for Network and Service Continuity in Wireless Mesh Networks*, Signals and Communication Technology, DOI: 10.1007/978-1-4614-4627-9_6, © Springer Science+Business Media New York 2013

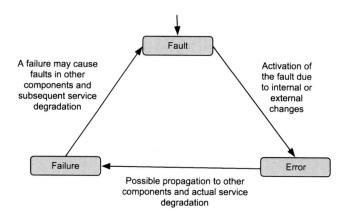

Fig. 6.1 Chain reaction fault, error and failure

as a failure (of a high-level component such as a link or a node). It is worth noting that the failure in turn can cause other faults in associated and dependent system components, and the chain continues with further possible degradation of service. This is shown in Fig. 6.1.

This articulation of fault, error and failure is useful in understanding service disruption metrics such as survivability and dependability. Even though their original definition may provide a way to quantify them, they have taken on more general meanings with time. The original definition of survivability is as below (Grover 2003):

Definition 6.1 *Survivability* is the ability of a network to continue providing service in an event of a failure.

It is a characterization of the time-varying performance of the network in the immediate aftermath of a fault until a new steady state is achieved (see Fig. 6.2). However, today, the term survivability typically signifies not only a quantitative measure of the system's transient ability to continue functioning in case of failure, but it is used more generally as an encompassing quality that includes many related measures. Similarly, dependability is also another general term that accommodates many quantitative measures including measures related to survivability and security. It is precisely defined by Algirdas Avizienis (2001) and Avizienis et al. (2004) as follows:

Definition 6.2 *Dependability* is the ability of a system to avoid failures that are more frequent or more severe, and outage durations that are longer, than is acceptable to the user.

Another definition commonly used for dependability (Al-Kuwaiti et al. 2009) is:

Definition 6.3 *Dependability* is the ability of a system to deliver service that can justifiably be trusted.

Fig. 6.2 Survivability quan-
tifies the network's transient
behavior after failure

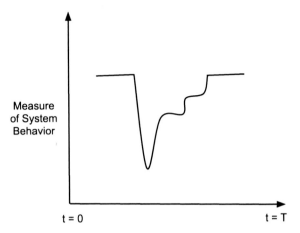

Note that in this definition, the term dependability is a network property that
encompasses survivability and security measures. Figure 6.3 attempts a classification
of various attributes.

6.1.1 Survivability Measures

Each of the survivability measures mentioned in Fig. 6.3 is explained next. A more
detailed description of these definition can be found in Haverkort et al. (2001).

6.1.1.1 Reliability

Distinct from its colloquial meaning, *reliability* of a system is technically defined as
a mission oriented question.

Definition 6.4 *Reliability* is the probability of a system performing its purpose ade-
quately for the period of time intended, under the operating conditions intended.

In other words, if a system started at the expected service level at time $t = 0$, and
the time of the mission is T, then reliability is the probability that the system will
not fail before the end of the mission. "What is the probability that the car engine
will work failure-free until the race is over?" provides an example of a reliability
question.

Formally, the reliability function can be defined as below:

$$R(t) = P\{\text{no failure in time}[0, t]\} \qquad (6.1)$$

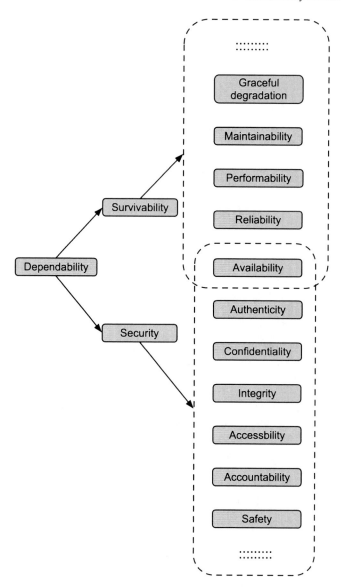

Fig. 6.3 A classification of continuity qualities of a system

Note that $R(t)$ is always a non-decreasing function with $R(0) = 1$ and $R(\inf) = 0$ (refer Fig. 6.4). *Cumulative Failure Function* $Q(t) = 1 - R(t)$ is the probability that there is at least one failure in the time interval $[0, t]$. In some systems as shown later, reliability is desirable to be defined as the complementary cumulative distribution function of time between failures. Let $f(t)$ be the failure density function that represents probability density function of time to next failure.

Fig. 6.4 Reliability is a
mission-oriented question

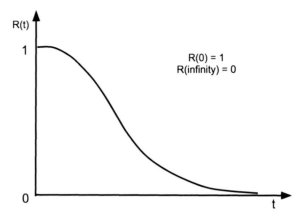

$$\int_{t_1}^{t_2} f(t)dt = \text{probability of at least one failure in time interval } [t_1, t_2] \qquad (6.2)$$

By definition of Cumulative Failure Function,

$$Q(t) = \int_0^t f(t)dt; \qquad (6.3)$$

and

$$R(t) = 1 - \int_0^t f(t)dt = \int_t^{inf} f(t)dt \qquad (6.4)$$

6.1.1.2 Availability

The mission-oriented nature of reliability is suitable for analyzing systems which should provide uninterrupted service for a predetermined duration. This is mainly because in such mission-critical systems, consideration of component repair and system restorability is unrealistic. For systems where expected service is almost perpetual in nature (such as the Internet), failures (and repairs) cannot be ignored and indeed may be frequent, and it is necessary that they are accommodated in the analysis. The measure suitable for such analysis is that of system availability.

Definition 6.5 *Availability* is the probability of finding the system in the operating state at any arbitrary time in the future.

Note that unlike reliability, we now consider systems that can be repaired. As an example, a question such as "What is the probability that a connection to a web server is available when my computer is turned on?" is an availability question. For repairable systems, it is also necessary to understand an additional attribute of maintainability.

Definition 6.6 *Maintainability* is the degree of ease by which a system can repaired in the event of failure.

It is clear that a system that exhibits high maintainability will offer a higher availability. Availability can be calculated using:

$$\text{Availability} = \frac{\text{Time for which the system was operational}}{\text{Total time of observation}} \quad (6.5)$$

The events of failures and repairs can be further understood using Fig. 6.5. The Figure also defines three terms (MTTF, MTBF and MTTR) which in turn can be used for extending Eq. 6.5 as below:

$$\text{Availability} = \frac{\text{MTTF}}{\text{MTTF} + \text{MTTR}} \quad (6.6)$$

Lastly, Fig. 6.6 shows a comparison between how availability and reliability changes with time when a system reaches a steady state. It shows that understanding of steady state availability holds a special importance in repairable systems such as most types of networks. In the light of our early discussion on the potential use of WMNs in Internet retrofit or municipal networks, and in replacing carrier-grade access networks, availability is indeed an important survivability attribute for mesh networks.

Network Availability

Network availability is typically understood with consideration of its high-level components such as nodes and links. For illustration purposes (refer to Fig. 6.7), let us consider a network where edges are prone to failure with a given probability (component availability), while the nodes do not fail. Any such network can be modeled using a random graph. Let $G(V, E)$ be the network graph with n nodes and m edges in the edge set $E = \{e_1, e_2, \cdots, e_m\}$, let $P = \{p_{e_1}, p_{e_2}, \cdots, p_{e_m}\}$ be the probabilities of their correct operation. Given these settings, a large number of availability problems can be analyzed for such a network. Three availability problems listed below have been widely studied in literature:

Definition 6.7 *2-terminal Availability* is the probability that two given nodes are connected to each other.

Definition 6.8 *Network Availability* (or connectivity) is the probability of all nodes being connected with each other.

Definition 6.9 *k-terminal Availability* is the probability that k given nodes are connected to each other.

All the above problems are known to be computationally expensive even for a small network size. This can be understood from the illustrative network of Fig. 6.7. Finding the probability that node A and node F are connected requires calculating a large polynomial of p.

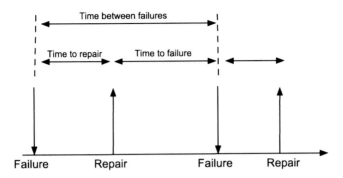

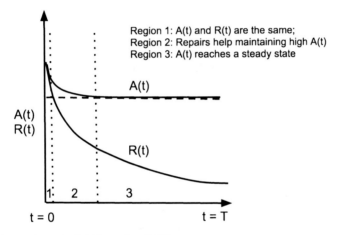

Fig. 6.5   Understanding availability in terms of MTTF, MTTR and MTBF

Fig. 6.6   Comparison of reliability and availability

Fig. 6.7   Graph representation
of a network where each link
can operate correctly with
probability of 0.9

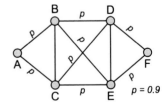

To understand this, first let us assume, for the moment, that all the edges in the
network operate with the same probability p. A *pathset* in context of any availability
problem is a set of edges that, when operational, guarantees connectivity between
the desired set of nodes (2 nodes, all nodes or k nodes). Similarly, a *cutset* is set of
edges that, when failed, causes disconnection among the desired set of nodes. Let

Table 6.1 Complexity of availability problems Colbourn (1987)

Problem	Graph	MinPath	MinCut
2-terminal	#P-complete	open	polynomial
Network	#P-complete	polynomial	open
k-terminal	#P-complete	open	open

N_i denote the number of pathsets with i edges. The availability now can be written as a polynomial in p:

$$Rel(G) = \sum_{i=0}^{m} N_i\, p^i (1 - p)^{m-i} \qquad (6.7)$$

The above availability polynomial is written in terms of pathsets, and similar polynomials can be written in terms of cutsets and complements of pathsets. Finding a sequence of coefficients $\{N_1, N_2, \cdots, N_m\}$ (or $\{C_1, C_2, \cdots, C_m\}$ for cutsets) is actually a problem of counting pathsets which in turn makes the availability evaluation problem a counting problem.

In case of 2-terminal availability, pathsets are all possible paths which connects the two nodes and cutsets are all possible edge-cuts which disconnects them. Similarly, in network availability, pathsets are all possible spanning trees and cutsets are all possible network cuts. Now, let l be the size of a minimum cardinality minpath (minimal pathset) and similarly let c be the size of a minimum cardinality mincut (minimal cutset). It was shown by Colbourn (1987) that if, for an availability problem, it is NP-hard to compute either l or c, the corresponding availability problem is also NP-hard. Similarly, if computing N_l or C_c (or any other coefficient) is #P-complete, computing the corresponding availability is also #P-complete. 2-terminal availability, all-terminal availability and k-terminal availability, all are proven to be #P-complete using the above, since computation of the coefficients are #P-complete in almost all cases.

For a network such as the one shown in Fig. 6.7 where edges can be in either working or failure mode, the complexity of the availability problem can be understood as follows. In a network with m edges where edges are either in failed or operational mode (and nodes do not fail), there are a total of 2^m network states. Depending on the desired set of nodes between which the connectivity is necessary, 2-terminal, network or k-terminal availability can be understood in terms of network states. Let C be the set of states where desired set of nodes are connected, and similarly, let D be the set of states where desired set of nodes are disconnected. In such a case, availability is defined as $\frac{|C|}{2^m}$, and similarly, unavailability can be defined as $\frac{|D|}{2^m}$ (Table. 6.1).

6.1.1.3 Performability

Availability is one of the many fundamental properties of any network, and it is widely studied in literature. However, there are certain limitations of availability metric. Availability can only distinguish between connected states and disconnected states of the network. Due to this characteristic, it cannot be applied to a performance degradable system. As an example, failure of a link might not cause a complete disconnection of the network but might affect the expected performance delivered by the network (such as throughput, delay etc.). The binary nature of availability is not sufficient to model such performance-varying states of the network. Many practical systems (especially networked systems) being performance degradable in nature, the *performability* metric has been defined and studied in literature (Haverkort et al. 2001) as a more appropriate metric.

Formally, let $S = \{s_{e_1}, s_{e_2}, \cdots, s_{e_m}\}$ describe a network state where $s_{e_i} = 1$ if edge e_i is operational and 0 otherwise. Let $\mathscr{X}$ be the set of all 2^m network states. Also, let $F(S)$ be a performance function defined on the state S. The probability that the system is in state S is given by $\Pr(S)$ (also referred to as the state occurrence probability). The well-known performability metric can then be calculated as

$$\bar{P} = \Sigma_{S \in \mathscr{X}} (F(S) \cdot \Pr(S)) \tag{6.8}$$

It is worth noting that availability can be regarded as a special case of the more general metric performability, obtained when the performance function is defined as a binary connectivity function, i.e. $F(S) = 1$ when the desired set of nodes are connected, and 0 otherwise. Also, performability permits statistical dependence among link failures (as opposed to availability).

The computation complexity of performability in turn involves complexities of availability and performance problems. Depending on the networked system at hand, obtaining its performance can be a feasible (e.g. wired networks with known traffic demands) or a hard problem (e.g. wireless networks). Note that in cases like the availability problem, $F(S)$ can be binary (S is connected or not) but $F(S)$ can take any form in general which makes the performability evaluation problem even more challenging. Its exact evaluation is known to be a hard problem (Harms 1995) and most of the current evaluation schemes depend on state generation methods.

6.1.2 Random Graphs

A random graph $G_p(V)$ is a graph defined on $V = v_1, v_2, \ldots, v_n$, and a link exists between a pair of nodes with the probability p. Random graphs have been largely used in modeling of wired networks where each link has a failure probability. One fundamental property of random graph is that failure of each link is independent of others. Since in wireless networks, the failures are likely to be distance dependent due to characteristics of RF propagation, the native definition of random graphs requires adaptation before using them to model wireless networks. To understand this, the characteristics of original random graphs are discussed next.

6.1.2.1 Node Degree

It was shown by Bela (2001) that the number of nodes connected to a node v_i in $G_p(V)$ follows a binomial distribution:

$$\Pr\{D_{v_i} = d\} = \frac{n-1}{d} p^d (1-p)^{n-1-d} \tag{6.9}$$

Let $d(v_i, v_j)$ be the distance between nodes n_i and n_j. When using a random graph to model a wireless network, the link existence/failure probability is in fact distance dependent. This allows us to extend the notation of random network graph $G_p(V)$ to $G_{p_{d(v_i,v_j)}}(V)$ for wireless networks. As defined by Hekmat and Van Mieghem (2003), the total number of edges (m) between nodes in such random graph can be calculated as below:

$$m = \sum_{i=1}^{n} \sum_{j=i+1}^{n} p_{d(v_i,v_j)} \tag{6.10}$$

Now if the nodes are uniformly distributed over the two-dimensional plane, Hekmat and Van Mieghem (2003) derived the expected number of links using the dissection method, where the network area is assumed to be covered with s small squares, and each of the squares is small enough to contain at most one node. Hekmat and Van Mieghem (2003) showed that the mean number of edges in $G_{p_{d(v_i,v_j)}}(V)$ can be given as:

$$E[m] = \frac{n(n-1)}{s(s-1)} \sum_{i=1}^{s} \sum_{j=i+1}^{s} p_{d(v_i,v_j)} \tag{6.11}$$

Similarly, Hekmat and Van Mieghem (2003) defined the link density to be the ratio of $E[m]$ to the maximum possible (undirected) number of links $n(n-1)/2$. This can be calculated as below:

$$\mathscr{M} = \frac{2\,E[m]}{n(n-1)} = \frac{2}{s(s-1)} \sum_{i=1}^{s} \sum_{j=i+1}^{s} p_{d(v_i,v_j)} \tag{6.12}$$

This further facilitates the calculation of the mean node degree as below:

$$E[D] = \frac{2E[m]}{n} = (n-1)\mathscr{M} \tag{6.13}$$

6.2 Topological Resilience of Wireless Mesh Networks

Wireless mesh networks have come a long way from the inception of the concept and their first deployment. Since the concept of WMNs was derived from the precursor ad-hoc networks model, their initial deployments followed the unplanned community driven model.

In an unplanned community wireless mesh (Bruno et al. 2005), members of the community host a mesh node in their own facility (homes, lightpoles etc.). Any such mesh node allows access not only to the host but also to the nearby neighborhood. Such a mesh node is also responsible for forwarding the data traffic of other mesh nodes, and this mutual contribution builds a wireless access network. The fundamental advantages of such networks are the capability of incremental expansion, and cheap deployment. A large number of deployments were carried out based on this unstructured design idea, many of which are still operational (see Bruno et al. 2005).

As the concept of wireless mesh networks matured, it became apparent that unstructured deployments are not suitable for municipal networks mainly due to two reasons. First, the incremental design did not allow enough opportunities for performance improvement, and second, it became increasingly difficult to stimulate participation in such networks especially when the network was unable to deliver the expected performance. This led to design and development of planned wireless mesh networks.

6.2.1 Component Failures

6.2.1.1 Node Failures

As opposed to ad-hoc networks where nodes are resource constrained (e.g. battery powered, limited buffer size etc.), nodes in mesh networks are generally not resource constrained. Especially in a carrier-grade mesh network, nodes are powered continuously using an uninterrupted power source. Also, the mesh nodes typically have enough processing power and memory to allow one or more radios to communicate continuously. This results in lower failure rate of the mesh nodes. Note that it is not possible to completely circumvent node failures because certain high level failures (such as software failures) can still occur. It is also observed in many commercial deployments that even in the case of node failure, the repair of node is relatively less time consuming. This is because in most current mesh systems mesh nodes are controlled remotely using a high capacity wireless controller that can resolve failures (especially software failures) automatically, and enables a faster repair process.

Due to this lower failure rate (lower MTTF) and faster repairs (lower MTTR), the availability of a mesh node is typically very high.

6.2.1.2 Link Failures

Links of wireless mesh networks are inherently susceptible to failures due to the unpredictability of the wireless medium. In urban environments especially, radio propagation can be highly non-uniform and coverage regions of mesh nodes can be quite non-circular, and even of nearly arbitrary shapes (Robinson et al. 2010) (see Fig. 6.8). This non-uniformity in radio propagation, which results into probabilistic connectivity between mesh nodes is known as *shadowing*. Due to shadowing, the signal power at different points with the same expected path loss shows significant variations. Prior measurement studies have shown that path loss at a particular location is random and distributed log-normally around the average distance dependent path loss. This makes the received signal strength inherently probabilistic. Formally, as shown by Rappaport (1996), for a transmitting node u, the received signal strength at receiver v at a distance d_{uv} is modeled as:

$$PR_{dBm}^{v}(d_{uv}) = PT_{dBm}^{u} - PL_{dB}(d_0) - 10\,\eta\,\log_{10}\left(\frac{d_{uv}}{d_0}\right) + S_{dB}^{\sigma} \qquad (6.14)$$

In the above, PT_{dBm}^{u} is the transmit power, $PL_{dB}(d_0)$ is the reference path loss at a distance d_0, η is the path loss exponent and S^{σ} is a zero-mean Gaussian variable with standard deviation σ. A wireless link exists between u and v if the received power $PR_{dBm}^{v}(d_{uv})$ is not less than some given minimum threshold P_{min}. Such a threshold is typically referred to as the *receiver sensitivity*. The distance d_r is the communication range in the absence of shadowing (i.e. $\sigma = 0$).

$$d_r = d_0\,10^{\frac{PT_{dBm} - PL_{dB}(d_0) - P_{min}}{10\,\eta}} \qquad (6.15)$$

An error function (or Q-function) can be used to determine the probability that the received signal power is larger than P_{min}. The probability ($\Pr(uv)$) that there exists a wireless link between u and v can be given by:

$$\Pr\{PR(d_{uv}) \geq P_{min}\} = \frac{1}{2}\left[1 - \mathrm{erf}\left(\frac{10}{\sqrt{2}} \cdot \log_{10}\frac{d_{uv}}{d_r} \cdot \frac{\eta}{\sigma}\right)\right] \qquad (6.16)$$

Typically state-of-the-art measurement and modeling methods (Robinson et al. 2008a) are used to derive the exact values of η and σ for a given outdoor environment. The value of d_{min} (along with receiver SINR) determines the data rate at which the established link will operate. Given the RF profile (η, σ and $PL_{dB}(d_0)$) of the environment, the mesh node profile (receiver sensitivity and corresponding data rates) and node positions, Eq. 6.16 can be used to determine the *shadowing probabilities* of the existence of links (or its complement, failure of links).

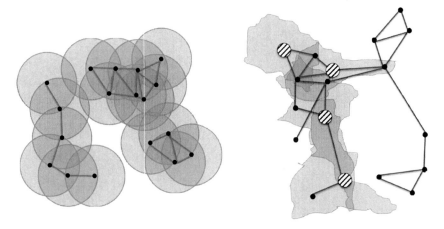

Fig. 6.8 Connectivity of the same set of node positions under disk-based propagation model (*left*: coverage disks shown for all nodes), and shadowing propagation model (*right*: coverage areas shown for highlighted nodes)

6.2.2 Topological Resilience

A mesh topology provides more redundant routing paths compared other network topologies such as a tree. Due to this reason, wireless mesh networks are especially robust to failures. In case a particular link is unavailable due to some reason, the data can be forwarded along alternate paths. This re-routing requires protocol-level assistance from networking layer, and such automatic re-route mechanisms are already built in most of the routing protocols discussed in Chap. 4.

Two fundamental topological factors, node placement and transmission power control, determine the topology of wireless mesh networks. Other factors such as receiver sensitivity or antenna propagation pattern also affect the topology which is discussed in subsequent sections. Given a node placement, the establishment of a link between any two nodes depends on the distance between them, the path loss exponents (η) and the shadowing coefficient (σ) as shown in Eq. 6.17. As shown by Hekmat and Van Mieghem (2003), Eq. 6.17 can be written as:

$$\Pr\{PR(d_{uv}) \geq P_{min}\} = \frac{1}{2}\left[1 - \mathrm{erf}\left(\frac{10}{\sqrt{2}} \cdot \log_{10}\frac{d_{uv}}{d_r} \cdot \frac{1}{\xi}\right)\right] \qquad (6.17)$$

where $\xi = \frac{\sigma}{\eta}$ is called the propagation ratio. ξ holds a special importance in topology establishment of a mesh network since both σ and η are RF characteristics of the neighborhood where the mesh network is being deployed. Note that when $\xi = 0$, the resultant propagation model is a path-loss model without shadowing effect. A high value of ξ means severe shadowing and lower path-loss, conditions which are especially common in urban neighborhoods where mesh networks are deployed. Note that typical value of η ranges from 2 (in free space) to 6, and value of σ can be

Fig. 6.9 The effect of shadowing on link existence: link
probability is nearly a step-
function for zero shadowing,
but shows a more gradual
change for increasing ratio of
shadowing and path loss

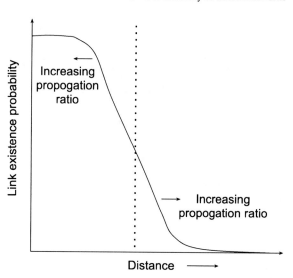

as high as 12 dB in severely shadow fading environments. This results into value of ξ that ranges from 0 to 6 Hekmat and Van Mieghem (2003).

For a given ξ, let us first consider the relation between link establishment probability (Eq. 6.17) and distance between two nodes. Its general behavior is shown in Fig. 6.9 with normalized distance $r = \frac{d_{uv}}{d_r}$. As expected, when $\xi = 0$, link probability does not change with variation in distance as the propagation model reduces to the path-loss model. For a higher value ξ, it is observed that link probability between nodes with lesser distance decreases, but link probability between nodes farther from each other increases. This phenomena is especially common in urban mesh networks where it is possible that a node is not connected to a nearby node but in fact can establish a link to another node that is farther away, as we previously illustrated in Fig. 6.8. Another control parameter that affects the shadowing phenomena is transmission power control. For a given value of ξ, increasing transmission power allows establishing links between nodes farther away from each other. Hekmat and Van Mieghem (2003) show that when the transmit power level of nodes increase, the distance upto which there is a non-zero probability of having a link also increases. The probability of link existence is thus related to various factors such as transmission power, node placement and RF profile of the environment. They also study the coverage area of nodes and how it changes with changing value of ξ in order to understand the increase of shadowing in terms of node degree of mesh nodes.

Substituting the value of link establishment probability from Eq. 6.12 into Eq. 6.17, link density in case of shadowed environment can be calculated as below:

$$\mathcal{M}_{sh} = \frac{1}{s(s-1)} \sum_{i=1}^{s} \sum_{j=i+1}^{s} \left[1 - \mathrm{erf}\left(\frac{10}{\sqrt{2}} \cdot \log_{10} \frac{d_{ij}}{d_r} \cdot \frac{1}{\xi} \right) \right] \qquad (6.18)$$

Thus increasing values of ξ have an increasing effect on link density. Mean node density can be used to calculate the value of average node degree using Eq. 6.13; Hekmat and Van Mieghem (2003) show that the degree distribution is roughly a unimodal one, and the mode of the distribution shifts to the right (higher degree) with increasing values of ξ.

6.3 Availability

In a planned mesh network deployment, rigorous analysis methods are used to understand RF profile of the neighborhood (methods such site survey), and coverage requirements are taken into consideration beforehand. These methods of informed design improved the performance of the networks, and the mesh networks were subsequently started belonging to the category of carrier-grade access networks. In this phase, mesh networks were being designed and deployed for better performance, but the performance remained fundamentally best-effort in nature, mostly because reliability and performance guarantees were not quite understood from the designer's perspective.

With the advancements in physical layer technologies in last few years such as smart antennas and MIMO, many recent deployments of wireless mesh network have shown that they can satisfy the ever increasing traffic demand of urban areas. Mesh networks also continue to be a suitable candidate for municipal networks because they have recently been shown to support numerous applications (GoogleWiFi 2007; Kansas 2011; Ponca 2007; Vos 2011) like meter reading, surveillance, emergency services and public safety, transportation assistance etc. Mesh networks are now expected to deliver carrier-grade services instead of best-effort services. With these increasing expectations, it is also generally being recognized that there is a lack of good understanding of the service consistency that mesh networks offer to its users. This is natural since ad-hoc networks (from which the mesh concept originated) were never designed for the carrier-grade high-performance, high-availability services.

The performance and reliability of contemporary wireless mesh network should be compared to their current contenders in access networks such as cellular network. This requires that novel analysis methods are developed, and networks are designed using them instead of adapting the understanding from ad-hoc networks research. But mesh networks must be analyzed in the same context as other contenders— for example, there are various methods for evaluating service continuity of cellular networks (e.g. Annamalai et al. 2001) in case of component failures.

Developing an exact evaluation of the constancy of service provided by a mesh network to its users in presence of link failures is a challenging problem. Distinct from purely network connectivity or QoS related approaches, it requires focus on metrics which are relevant in providing continuous service to users. The current mesh network user expects an "always-on" attribute in service where total disruption is seen as a serious failure of network, while small variations does not count as serious shortcomings. Failures being unavoidable from the designer's perspective, it is necessary that more pragmatic survivability metrics are developed for mesh networks that can be used for service level agreements. We identify two fault tolerance metrics and adapt their native definitions to urban-scale mesh networks. The first appropriate metric is availability; as we mentioned above, it measures the fraction of time (over a long period) a given system is in the working state; i.e. the probability of finding it working at a random inspection. Since the user does not care how the mesh provides connectivity to Internet, we formulate the availability problem as a k-*center availability* problem. This is because the Internet gateway nodes of mesh networks are ideally referred as the centers of the network connectivity graph. The k-center availability[1] problem is to find the probability that every mesh node is connected to at least one of the k Internet gateways.

Having articulated the most important concern, connectivity, we turn to the next metric, that of network throughput performance experienced by users. As we remarked, users do not penalize the network for short term variations in the throughput experience, but over longer periods expect a reasonable aggregate experience. A network that frequently lapses in providing expected throughput will eventually be abandoned by users. The appropriate metric for this is that of performability which is the time integral of a performance metric (here, throughput) as described earlier. Performability not only accounts for the performance delivered by the network, but more importantly considers how consistently this performance is delivered. We believe that k-center performability is an important metric to consider in understanding the service continuity experienced by mesh users.

In sections that follow, we define k-center availability (KCA) and performability (KCP) metrics. Using different state generation methods, we then develop algorithms for their exact evaluation. For KCA, we describe a constrained Monte Carlo state generation method which utilizes bounds derived from network graph to reduce the number of states required by the Monte Carlo simulation. In case of KCP, we describe an approach of applying the most probable states generation method to a subset of edges which are crucial in determining the performability. One of the important advantages of these methods is that they can be applied to moderate to large mesh networks with as many as 500 nodes. Next, equipped with these evaluation methods, we analyze the impact of two basic topology formation factors—node density and transmission power—on KCA and KCP of mesh networks. This study leads to various interesting observations which can be further used for designing networks with high KCA and KCP values.

Unlike ad-hoc networks, links are more vulnerable to failures as compared to nodes in mesh network. This is mainly because the nodes in carrier-grade mesh networks are typically not power constrained (e.g. battery operated) or resource constrained.

Formally, let $G_P(V, E, P)$ be a the network graph of a mesh network. let n be the total number of mesh nodes and m be the total number of edges. For the edge

---

[1] This or similar metrics have sometimes been referred to as k-*center reliability* in recent literature such as Lee et al. (2011), but here we use the term availability in keeping with accepted definitions of these terms as referred to earlier.

set $E = \{e_1, e_2, \cdots, e_m\}$, let $P = \{p_{e_1}, p_{e_2}, \cdots, p_{e_m}\}$ be the probabilities of their correct operation. We refer to G_P as a probabilistic connectivity graph. The failure probabilities can be determined by any method which can model link failures of the given network. To focus on urban mesh networks, we restrict our attention to shadowing probabilities here. The probabilistic connectivity graph decouples the KCA and KCP evaluation methods (presented below) from how the link failure probabilities are determined. The evaluation methods can be applied to any set of edge failure probabilities given that the link failures are independent.

Given a mesh network where links can fail due to shadowing or any other reason, defining and calculating availability is a challenging problem. We next define mesh network availability:

Definition 6.10 *Mesh Network Availability* is the probability that every mesh node is connected to at least one gateway.

Since the gateway nodes provide Internet connectivity, the above definition argues that when a mesh network is available, all mesh nodes are in fact connected to Internet. This also guarantees that all mesh clients connected to mesh access points have the access to Internet services.

Even though the definition is simplistic in nature, accurate calculation of availability is indeed difficult. This is because the availability definition is different from conventional 2-terminal availability or network availability. Solutions to the 2-terminal availability can be applied to mesh network given that every mesh node has a fixed pre-assigned gateway. In case of a failure, if the mesh node is not connected to its gateway, the mesh node is denoted to be disconnected. Using this argument, the problem can be formalized as below.

For G_P, let $K \subset V$ be a set of k gateways where $K = \{g_1, g_2, \ldots, g_k\}$. Now, k-center reliability is defined as the probability that $\forall u \in V - K$ is connected to at least one gateway $g \in K$. Let $g(v_i)$ denote the gateway assigned to node v_i, and $Pr(v_i, v_j)$ be the probability that nodes v_i and v_j are connected (2-terminal availability). Now, the mesh network availability A_{mesh} can be calculated as:

$$A_{mesh} = \Pi_{i=0}^{n} Pr(v_i, g(v_i)) \qquad (6.19)$$

There are multiple disadvantages of representing mesh availability in form of 2-terminal availability. First, the complexity of 2-terminal availability problem itself is very high, and the above representation increases the complexity by the factor of $n - k$. Second and more importantly, restricting a mesh node to connect to only one gateway does not take full advantage of underlying robustness of a mesh topology. It is better if a mesh node disconnected from its nearest gateway is able to connect to another gateway to forward its data.

Similarly, network availability is a conservative estimation of mesh availability since there can be a large number of network states in which every mesh node is connected to a gateway but in fact all nodes of the network are not connected with each other (shown in Fig. 6.10). Even though this is true, network availability can in fact act as a valid lower bound on mesh availability.

6.3.1 802.11s Mesh Network

Among various efforts to calculate mesh availability, Egeland and Engelstad (2009) analyzed it as a k-terminal availability problem. In 802.11s mesh standard, the following terminology is used. *Mesh Point* (MP) is a mesh node in a mesh network (Fig. 6.11).

Definition 6.11 *Mesh Access Point* (MAP) is an MP that includes the functionality of an 802.11 access point, and can provide access to clients.

Definition 6.12 *Mesh Portal Point* (MPP) is an MP that includes the functionality of connecting the mesh network to Internet or other networks.

Egeland and Engelstad (2009) analyse a simple network topology where a boundary of MP and MAPs are linearly connected to an MPP, and this linear network surrounds a variable core mesh topology. If all $n-1$ distribution nodes (MP or MAP) are connected to the root node (MPP), a mesh network is in a connected/available state. This allows formulating the mesh availability problem as a k-terminal reliability problem where k terminals of interests are $n-1$ distribution nodes and the root node. This way, mesh availability can be calculated as follows:

$$A_{mesh} = 1 - \sum_{i=\beta}^{m} C_i^{r,d_1,d_2,\ldots,d_{n-1}} p^i (1-p)^{m-i} \qquad (6.20)$$

where p_d is the probability of link failure, $C_i^{r,d_1,d_2,\ldots,d_{n-1}}$ denotes the number of edge cutsets with exactly i edges, m is the number of edges, and β is the edge connectivity of the graph. This shows that when a cutset with number of edges between β to m occurs, the k nodes of interests are not connected to each other.

Egeland and Engelstad (2009) show that adding redundant nodes to the core topology facilitates additional links, and improves network connectivity. Increase in link failure probability causes sharper decrease in mesh availability (due to polynomial relation between them as in Eq. 6.20). For the given topology, when the link failure probability is higher than 0.6, the network can be denoted disconnected and dysfunctional, even with added node redundancy. As more and more redundant nodes are added to the topology, the resultant mesh availability increases. It is worth noting that adding redundant nodes can increase the network edge connectivity and sharper increase of mesh availability.

Such addition of redundant nodes leads to variety of optimization problems in which the choice of number of redundant nodes and their placement should be minimize the cost of additional nodes while reaching the expected mesh availability. This is further motivated by an alternate topology shown by Egeland and Engelstad (2009). In the alternate topology, the redundant nodes are added randomly instead of careful placement. Because the addition of the nodes are random, the link establishment is not controlled, and it is difficult to achieve a higher edge connectivity. This

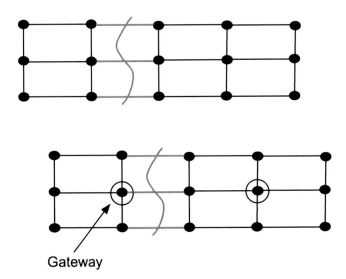

Gateway

Fig. 6.10 Mesh network connectivity is different from network connectivity

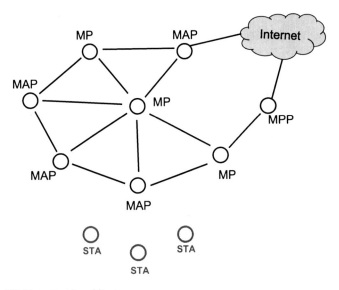

Fig. 6.11 802.11s network architecture

results into slower increase of mesh availability with addition of redundant mesh nodes.

One difficulty in using Eq. 6.20 for calculating mesh availability is that the calculation of number of cutsets with given number of edges is computationally expensive (as remarked in Sect. 6.1.1.2) in larger networks. This restricts the applicability of the approach to smaller mesh networks, and the approach is not efficient in calculating availability of large urban-scale mesh networks (such as GoogleWiFi (2007) or Ponca (2007) City Mesh).

6.3.2 Approximating Mesh Availability

The difficulty in exact evaluation of availability of mesh networks has led to many research attempts in which the availability is approximated using a computationally efficient solution. As was shown in the last section, cutsets based method introduces an inherent difficulty in calculating mesh availability due to complexity of finding cutsets with fixed cardinality. Camp et al. (2006) presented an approach in which all possible paths from every mesh node to all gateways are found, and the mesh availability is determined using the fact that when at least one of such path is operational for all mesh nodes, the mesh network is denoted available. The procedure to find mesh availability Camp et al. (2006) is as follows.

- Let V be the set of n nodes and E be the set of m edges
- Let $K \subset V$ be a set of k gateways where $K = \{g_1, g_2, \ldots, g_k\}$
- Let $V - K = \{v_1, v_2, \ldots, v_{n-k}\}$ be the set of mesh nodes
- For the edge set $E = \{e_1, e_2, \cdots, e_m\}$, let $P = \{p_{e_1}, p_{e_2}, \cdots, p_{e_m}\}$ be the probabilities of their correct operation
- For each wireless mesh node $v_i \in V - K$, find the set of all possible routes from v_i to all $g_k \in K$. Let R_{v_i} be the set of routes for node v_i
- For a node v_i, let $P(R_{v_i})$ be the probability that at least one path in R_{v_i} is operational
- The *average* mesh availability can be calculated as $A_{mesh} = \frac{1}{n-k} \sum_{i=1}^{n-k} P(R_{v_i})$

Note that the calculated availability value here is an average value and not an exact calculation. Camp et al. (2006) applied the method for availability evaluation to a square grid topology with varying level of gateway density.

Definition 6.13 *Gateway Density* (or *wire ratio*) of a mesh network is the ratio of the number of gateway nodes to the number of mesh nodes in a mesh network.

Camp et al. (2006) investigated how mesh availability changes with decrease of mesh node density for a given gateway density. As expected, the availability falls with increasing distance between the mesh nodes; the curve has a general sigmoid nature, so that the availability is most sensitive to distance in the middle section. The onset of the reduction occurs earlier for smaller wire ratios. This indicates an opportunity to obtain a favorable tradeoff between availability and cost, by choosing the wire ratio and distance between nodes to minimize cost as much as possible while the availability is still comparatively insensitive to it. For the RF parameters under consideration, an inter-node distance of 200 m was found to yield the highest possible availability with least total node and gateway cost.

Placement of gateway nodes in turn motivates many different design problems because there might be only a few geographical locations where a wired connection

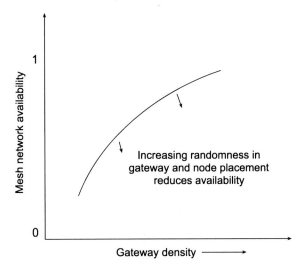

Fig. 6.12 Comparison of availability when nodes and/or gateways are randomly distributed

is available. Typically, the cost associated with assigning a mesh node the role of a gateway increases the cost of deployment due to installation and (as above) maintenance of wired connection. For a higher mesh node density, target availability can be achieved with fewer gateway nodes. This might be an efficient solution especially in deployments where only a few geographical locations are capable of hosting a gateway node.

Placing all the mesh nodes in a grid is usually not possible due to constraints on geographic locations and coverage requirements. Such constraints may be completely external to networking considerations, such as infrastructure logistics, legal, or human usage patterns. This motivates studying deployments where mesh nodes are randomly placed. A completely random placement (using Poisson point process) is likely to result into decrease of availability compared to a grid placement. This is mainly because any perturbation from grid topology can not increase the network-wide connectivity. This requires a designer to study the availability question where mesh nodes and gateways are randomly placed in the network area. Camp et al. (2006) compared the availability of mesh network in three separate cases (i) mesh nodes and gateways are placed in square grid, (ii) mesh nodes are placed in square grid but gateways are randomly distributed, and (iii) mesh nodes and gateways are randomly distributed. Their results show that when gateways are placed in a grid form, the expected availability can increase by as much as 20 % (Fig. 6.12).

The problem of node and gateway placement in urban-scale mesh networks can be slightly different due to characteristics of urban neighborhoods. In such cases, it might be possible to in fact place the gateway nodes in a grid fashion because of availability of a large number of locations (e.g. office buildings, commercial centers etc.) where gateways can be placed. On the other hand, placement of the mesh nodes might be more constrained due to coverage requirements. Investigations by Camp et al. (2006) confirm that increased randomness in either node placement, or gateway

Fig. 6.13 Node and gateway placement of Ponca (2007) City Mesh network

placement, or both, tend to progressively reduce availability. In Ponca (2007) City Mesh (Fig. 6.13), it can be observed that the mesh nodes are nearly randomly placed but the gateways in fact are placed nearly in a grid form. Studying the availability of such cases is still an open research problem that requires optimizing a large number of parameters.

6.3.3 Constrained Monte Carlo Simulation Approach to KCA

In this section, we provide a constrained Monte Carlo simulation method for the exact evaluation of KCA. We then analyze the impact of two topology formation factors, node density and transmission power, on KCA. Also, when all non-gateway nodes can reach at least one of the gateways, we mark the network to be *k-center connected*.

As distinct from various other approaches in literature which focuses on deriving asymptotic results (Bettstetter and Hartmann 2005; Hekmat and Van Mieghem 2003; Xing and Wang 2008) for network connectivity, we are interested in the exact evaluation of KCA. Also, network reliability can not be used for estimating KCA in any way because there can be a large number of network states where network is disconnected due to a link failure but the it is still k-center connected. Other reliability metrics are widely studied in network reliability literature (Colbourn 1987) but they are not useful in finding KCA. The most general form of reliability problem is that of finding t-terminal reliability which is the probability that the given t nodes are connected. The problem is known to be #P-complete (Colbourn 1987) for $t = 2$ (2-terminal reliability) and $t = n$ (network reliability). Due to this, most of the current methods of finding 2-terminal and network reliability rely on Monte Carlo state generation.

A network of m edges where each edge can be in either operational or failed state has a total of 2^m possible network states. Since the focus of this study is to find availability of mesh networks with as many as 500 nodes, exact evaluation methods (Colbourn 1987) such as complete state enumeration, graph transformation, factoring etc. are not useful. Such computational complexity has given rise to two approaches of calculating the availability. First, a set of reasonable upper and lower bounds are derived based on the graph structure and edge probabilities. In most cases, the bounds are computationally inexpensive to derive but their quality depends on various factors, and they are often too loose to be useful. Second, Monte Carlo simulations are used to estimate the exact value of availability. In this method, a pre-calculated number of network states are uniformly chosen from all possible states. The actual availability value is then an estimate based on the availability measures of these states. Although this allows exact computation of the availability value, the number of states necessary to guarantee the desired accuracy of estimate can be very high.

Fishman first proposed a reliability evaluation scheme Fishman (1986) (for 2-terminal reliability) which combines benefits of both these approaches. He suggested that even if the bounds found are only reasonably good, they can be used to direct a Monte Carlo simulation to reduce the number of necessary samples. This means that bounds should be used to eliminate a large number of states from the entire state space, and Monte Carlo simulation is then necessary to be run only on the remaining states. This is shown in Fig. 6.14. With the bounds, it is already known that all the states above the upper bound are connected and all the states below the lower bound are disconnected. Now, if we only generate the states in between the bounds (undetermined states) in our Monte Carlo simulation, fewer states will be necessary as compared to a naive Monte Carlo simulation. We refer to this scheme as a *Constrained Monte Carlo Simulation* (CMCS).

Even though the CMCS method seems to be an attractive choice for KCA evaluation, it should be noted that there are multiple challenges in adapting this scheme for mesh networks. The first challenge is to derive a *reasonable* set of bounds which can be used in CMCS. This implies that bounds derived in terms of k-center connected and disconnected states should be tight enough to be useful, otherwise the CMCS method degenerates to a naive Monte Carlo scheme. Also, since we want to study KCA while varying topological factors like node density and transmission power level, the bounds should remain reasonable in all cases.

The second challenge is that the CMCS method, initially proposed for 2-terminal reliability, must be adapted for KCA. This requires that the characteristics of conditional state generation (explained below) should be maintained even when states are being generated for KCA evaluation. We next describe how bounds can be derived, and present the CMCS method for KCA evaluation.

6.3.3.1 Edge-Packing Bounds

Let $S = \{s_{e_1}, s_{e_2}, \cdots, s_{e_m}\}$ describe a network state where $s_{e_i} = 1$ if edge e_i is operational and 0 otherwise. The state space with total of 2^m states is divided into two disjoint and exhaustive subsets $\mathscr{C}$ and $\mathscr{D}$. As shown in Fig. 6.14, $\mathscr{C}$ is a set of all operational (k-center connected) states while $\mathscr{D}$ is a set of all failed (k-center disconnected) sets. We use edge-packing bounds which are described next.

In a graph $G = (V, E)$, an *edge packing* (Colbourn 1987) of G by k graphs $G_1, G_2, \cdots, G_K$ is obtained by partitioning the edge set E into some $k + 1$ subsets $E_1, E_2, \cdots, E_k, U$ and defining $G_i = (V, E_i)$. This way, an edge-packing of a graph is a collection of edge-disjoint subgraphs of the graph. An edge-packing lower bound on the reliability can be obtained by finding a set of *edge-disjoint minimal paths* (edge-packing of minpaths), and obtaining the probability that all the edges of at least one of these paths operate correctly. Similarly, an upper bound on the reliability can be obtained by finding a set of *edge-disjoint minimal cuts* (edge-packing of cuts) and finding the probability that all the edges of at least one of these cuts fail. The edge-disjointness ensures that failure of every path (or occurrence of any cut) is independent from others since failure of an edge will only impact one path (or cut) at most.

For KCA, edge-packing of minimal pathsets is a set of edge-disjoint forests where in each forest, every tree contains a gateway and all mesh nodes belong to a tree. Let I be the total number of such forests which are denoted by $F_1, F_2, \cdots, F_I$. Each forest is a union of a minimal path from every mesh node to a gateway. If all edges of at least one of these forests are operational, every mesh node can reach at least one gateway and the resultant state is operational. Let Ω_1 be the set of states where there exists at least one operational forest. By definition, Ω_1 is a subset of $\mathscr{C}$ as shown in Fig. 6.14. Hence, a *lower bound B* on k-center availability is the probability that all the edges of at least one of these forests operate correctly.

$$B = 1 - \prod_{1 \leq j \leq I} \left(1 - \prod_{e_i \in F_j} p_i\right) \qquad (6.21)$$

Finding I edge disjoint forests where trees in each forest are rooted at gateways turns out to be a non-trivial task. This is because as per the Eq. 6.21, a higher number of such edge disjoint forests (maximizing I) yields a better bound. Also, the lower bound improves if each forest contains edges which are more reliable (higher operational probability) compared to other non-forest edges.

To find this edge packing of forests, we first determine the weight of every edge using $w_{e_i} = -\ln(p_{e_i})$. Now, let $d(u, v)$ be the total weight of all edges on the shortest

Fig. 6.14 State space for constrained Monte Carlo simulation

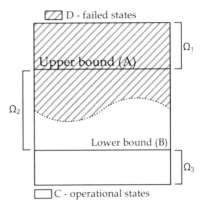

path between u and v. We then partition G_P into subgraphs $G_{g_1}, G_{g_2}, \cdots, G_{g_k}$ where $v \in G_{g_i}$ if $d(v, g_i) = \min\{d(v, g_j), g_j \in K\}$, and $e_i \in G_{g_i}$ if and only if both endpoints of e_i belong to the same subgraph. Let $\kappa(G_{g_i})$ be the edge connectivity of G_{g_i} and let $\kappa(G_P) = \min\{\kappa(G_{g_i}), g_i \in K\}$. Now from Tutte's theorem (Tutte 1961), $I = \lfloor \kappa(G_P)/2 \rfloor$ and G_P contains at least I edge disjoint forests where each forest contains a tree belonging every subgraph G_{g_i}. Once we have determined I, we can find these edge disjoint forests using a method proposed by Roskind and Tarjan (1985). To find I forests in which every node can reach a gateway using fewer and more reliable edges, we first find a score cw_{e_i} of every edge in a G_{g_i} as follows: $\forall e_i \in G_{g_i}$ between u and v, let $cw_{e_i} = \min\{d(u, g_i), d(v, g_i)\} + w_{e_i}$. Now, we sort all $e_i \in G_{g_i}$ in increasing order of their cw_{e_i} scores, and input the sorted list to the matroid partition algorithm of Roskind and Tarjan (1985) which then finds I edge disjoint trees in each G_{g_i} forming I edge disjoint forests.

Next we turn out attention to finding an upper bound A on KCA. There can be many different ways of determining a set of minimal cuts which can cause a failed network state. For any node, the set of all its incident links is a minimal cut. We observe that all such cuts corresponding to the nodes of an independent set of the graph are also edge-disjoint. Since any such cut disconnects a node from the rest of the network, such a node cannot be connected with any of the gateways; thus this is a failed state by definition. Let J be the total number of such cuts which are denoted by $C_1, C_2, \cdots, C_J$. If all the edges of at least one of these cuts are failed, the resultant network state is failed. Let $\Omega_3 \subset \mathscr{D}$ be the set of states where there exists at least one failed cut. An *upper bound* A on k-center availability is given by the probability that all edges of at least one of these cuts have failed.

$$A = \prod_{1 \leq j \leq J} \left(1 - \prod_{e_i \in C_j} (1 - p_i) \right) \qquad (6.22)$$

We use the well-known greedy algorithm for finding a maximal independent set, and then obtain the cuts as above.

6.3.3.2 Sampling Algorithm:

As in Fig. 6.14, since we already know that all states in Ω_1 are connected and all states in Ω_3 are disconnected, only the states in between the upper bound and lower bound are undetermined, and Monte Carlo simulations are required to determine their status. Let Ω_2 be the set of all such states. Note that Ω_2 is the set of states such that every cut C_j where $1 \leq j \leq J$ is operational and every forest F_i where $1 \leq i \leq I$ has failed. Since all cuts C_j are operational, there *can* be an operational path from every mesh node to a gateway; but if so, the union of all such paths is not a forest among F_i.

The sampling algorithm is presented in Algorithm 1 which is adapted from Fishman (1986), Manzi et al. (2001) for KCA. States of Ω_2 can not be uniformly sampled as in the naive approach because status of edges in a cut of a forest is conditionally dependent on each other. As described by Manzi et al. (2001), when sampling a state in Ω_2, the edges which belong to a forest and/or a cut have to be sampled sequentially (Procedure sampleState, Part 1). If an edge belongs to a fo rest, its status depends on the constraint that requires at least one edge of the forest to fail. This ensures that the forest is not operational. Similarly, if an edge belongs to a cut, its status depends on the constraint that requires at least one edge of the cut to be operational. This ensures that the cut has not failed. The edges not belonging to a forest or a cut can be sampled independently (Procedure sampleState, Part 2). The number of states required to be sampled from Ω_2 depends on the level of accuracy needed in KCA estimate. We use the standard deviation of KCA estimate as a guideline. Since the standard deviation decreases as more and more samples are generated, its value can be calculated after every fixed number of samples. If the targeted accuracy is achieved, no more samples are necessary and the procedure can be terminated with confidence.

6.3.3.3 Impact of Node Density and Transmission Power

In this section, we apply the CMCS method to study KCA of mesh networks with different topological configurations. We study two basic topology formation factors— node density and transmission power level to understand their impact on KCA of mesh networks. We assume that mesh nodes are distributed uniformly and all mesh nodes operate on a common transmission power level. It is also assumed that nodes are equipped with an omni-directional antenna for back-haul links (links between mesh nodes) and hence, Eq. 6.17 can be used to determine the shadowing probabilities. Unless mentioned explicitly, all physical layer parameters used in simulations are according to 802.11a standard. We choose node density and transmission power level for studying their impact on KCA because from a designer's perspective, they are two of the most important deployment decisions which affect both cost and throughput capacity.

To choose the RF profile of a real-world urban neighborhood, we use RF parameters of the GoogleWiFi network that were measured and studied by Robinson and Knightly (2007) and Robinson et al. (2008a). The network has a path-loss coeffi-

Algorithm 1: CMCS procedure to calculate KCR

Purpose: To calculate k-center reliability (KCR)
Input :
  Graph $G_P(V, E, P), n = |V|, m = |E|$,
  Set $K \subset V$ of k gateways, $K = \{g_1, g_2, \cdots, g_k\}$,
  $P = \{p_{e_i}, \forall e_i \in E\}$ probabilities of correct operation;

  Edge set of I forests $F_1, F_2, \cdots F_I$ and
  Lower bound $B = 1 - \prod_{1 \le j \le I}(1 - \prod_{e_i \in F_j} p_{e_i})$,
  $H_1 = \bigcup_{j=1}^{I} F_j$;

  Edge set of J cuts $C_1, C_2, \cdots C_J$ and
  Upper bound $A = \prod_{1 \le j \le J}(1 - \prod_{e_i \in C_j}(1 - p_{e_i}))$,
  $H_2 = \bigcup_{j=1}^{J} C_j, H = H_1 \cup H_2$;

  For $1 \le i \le m$ and $1 \le k \le I$,
  $Forest[e_i] = k$ if $e_i \in F_k$ and 0 otherwise,
  $\lambda[0] = 0$ and $\lambda[k] = \prod_{e_i \in F_k} p_{e_i}$;

  For $1 \le i \le m$ and $1 \le k \le J$,
  $Cut[e_i] = k$ if $e_i \in C_k$ and 0 otherwise,
  $\omega[0] = 0$ and $\omega[k] = \prod_{e_i \in C_k}(1 - p_{e_i})$;

  Accuracy threshold α;

Output : k-center reliability (KCR)
Method :

  $S = 0$;
  while $stddev(kcr) \le \alpha$ **do**
  | $X = X + 1$;
  | Generate a sample state using SAMPLESTATE;
  | Check if the sample state is k-center connected;
  | **if** *k-center connected* **then**
  | | $S = S + 1$;
  | **end**
  | $stddev(kcr) = \sqrt{\frac{(A-B)^2(1-S/X))(S/X)}{(X-1)}}$
  end
  Compute the k-center reliability as follows $kcr = B + (A - B)\frac{S}{X}$;
  return kcr.

PROCEDURE SAMPLESTATE:

  `// Sample edges in H - Part 1`
  while $\lambda[k] \ne 0$ *for* $1 \le k \le I$ and
  $\omega[j] \ne 0$ *for* $1 \le j \le J$ **do**
  | Select an edge e_i from a F_k or a C_j with fewest remaining edges and $\lambda[k] \ne 0, \omega[j] \ne 0$;
  | $a = Forest[e_i], b = Cut[e_i]$;
  | $p_{e_i}^* = \frac{p_{e_i} - \lambda[a]}{1 - \omega[b] - \lambda[a]}$;
  | Sample u from uniform distribution $U[0, 1]$;
  | $s_{e_i} = \lfloor u + p_{e_i}^* \rfloor$;
  | Set $\lambda[a] = \frac{s_{e_i} \cdot \lambda[a]}{p_{e_i}}, \omega[b] = \frac{(1-s_{e_i}) \cdot \omega[b]}{1 - p_{e_i}}$;
  end

  `// Sample edges in E - H - Part 2`
  forall the $e_i \in E - H$ **do**
  | Sample u from $U[0, 1]$;
  | $s_{e_i} = \lfloor u + p_{e_i}^* \rfloor$;
  end
  return the generated network state $(s_{e_1}, s_{e_2}, \cdots, s_{e_m})$;

Fig. 6.15 Increase of transmit
power level increases availability differently for different
densities

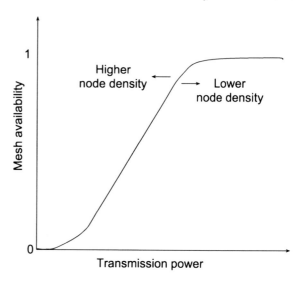

cient ($\eta = 3.5$), shadowing factor ($\sigma = 8\,$dB) and reference path loss of 50 dB at
a distance of 10 m. We also choose the receiver sensitivity to be $-80\,$dBm. Given
these parameters and transmission power level, the shadowing probabilities of links
can be calculated using Eq. 6.17. We then apply the CMCS procedure to calculate
the KCA. Figure 6.15 shows how KCA changes for a given node density as the common transmit power level is increased. As expected, KCA increases with increase in
transmission power because the network becomes more connected and the existence
probabilities of edges increase. Note that the increase in KCA is slower in the cases
of lower density which shows that high transmission power level is necessary in order
to maintain a high KCA in sparse deployments. That is, the phase transition width
(difference between maximum power level at which KCA is 0 and minimum power
level at which KCA is maximum) decreases quickly with increase in node density.

An optimization problem of designing a mesh network with guaranteed KCA
while utilizing minimum number of mesh nodes can be better understood using
Fig. 6.16. It shows the variation, with power level, of the minimum necessary node
density required to achieve a target level of KCA. Interestingly, increasing the target
KCA value from 0.99 to 0.999 (or 0.999 to 0.9999) requires significantly higher
node density or transmit power. This demonstrates that KCA evaluation is critical in
designing and provisioning a mesh network.

As in 802.11 standards, a higher value of receiver sensitivity ensures higher data
rate when the link operates in absence of interference. On the other hand, for a given
transmission power level, higher receiver sensitivity results in lower link existence
probabilities as per Eq. 6.17. To understand the impact of receiver sensitivity on
KCA, we fix the node density at 35 nodes per km^2, and vary the receiver sensitivity.
The results are shown in Fig. 6.17. It is observed that phase transition widths increase
with the increase in receiver sensitivity. The effect of higher receiver sensitivity on

Fig. 6.16 Minimum necessary transmit power—density characteristic for a given KCA moves toward increasing power and density with higher target KCA

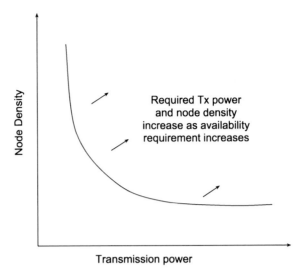

Fig. 6.17 Availability goes through phase transition from zero to one as transmission power increases

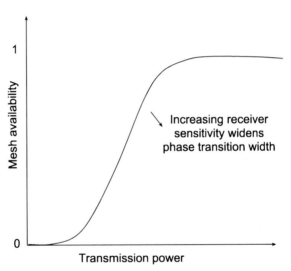

topology formation is that it reduces the link existence probabilities and the topology becomes sparser in general. We revisit the issue of link data rates and throughput capacity in light of k-center performability in Sect. 6.4.2.3. Note that in simulations, to distill the impact of changes in node density, we do not increase the number of gateways when increasing node density. Higher KCA values can obviously also be achieved using other design decisions such as increasing the number of gateways, using multiple radios and/or antennas at each mesh node, etc., which we do not discuss here.

6.3.3.4 Performance of CMCS Method

Figure 6.18a shows the nature of edge-packing bounds and the actual KCA value when node density is 15 nodes per km^2. Note that these bounds indicate the amount of reduction in search space when executing the Monte Carlo simulation. The percentage reduction can be calculated as $\pi = (1 - (A - B)) \times 100$. When $\pi = 0$, the CMCS method degenerates to a naive Monte Carlo method which is known to be inefficient for large networks. This is reflected in Fig. 6.18b which shows the time taken to estimate KCA with the accuracy of 10^{-5}. The simulations were run on a desktop computer with 1.6 GHz processor and 1 gigabytes of memory. Figure 6.18c shows the number of samples required to be generated for estimating KCA. As expected, simulation time and number of samples is proportional to π, and when the value of π is maximum, the corresponding value of number of samples is also the largest which leads to the longest simulation time. As we can see, the value of π is low in most cases, and the CMCS procedure results into substantial reduction in required number of samples over the naive Monte Carlo method.

6.4 Performability

K-center availability can be used by a network designer to understand what is the likelihood that the network will remain k-center connected in case of link failures. For any given state of the network, availability yields a zero-one evaluation of whether all the mesh nodes are connected to a gateway or not. Though this evaluation is absolutely crucial in providing highly available service to the users, it gives little or no information about the actual user experience. Even in a k-center disconnected state, users of mesh nodes which are connected to a gateway can be served successfully by the network. This requires evaluating the performance of a degradable system instead of denoting it to be operational or failed. Since link operational probability is essentially the availability of the link, it is necessary to quantify the difference between a high data rate link which has a very low availability and a moderate data rate link which is operational with high availability. The peformability metric is typically used for such a purpose in system survivability studies. It analyzes the performance of a failure-prone degradable network and evaluates how reliably the performance will be delivered by the network.

First we provide the definition of performability and then show how it can be adapted for the mesh network. As in KCA, let $S = \{s_{e_1}, s_{e_2}, \ldots, s_{e_m}\}$ describe a network state where $s_{e_i} = 1$ if edge e_i is operational and 0 otherwise. Let $\mathscr{X}$ be the set of all 2^m network states. Also, let $F(S)$ be a performance function defined on the state S. The probability that the system is in state S is given by $\Pr(S)$ (also referred to as state occurrence probability). The well-known performability metric can be calculated as $\bar{P} = \Sigma_{S \in \mathscr{X}} (F(S) \cdot \Pr(S))$. Note that in cases like KCA problem, $F(S)$ can be binary (S is k-center connected or not) but $F(S)$ can take any form in general which makes the performability evaluation problem even more challenging.

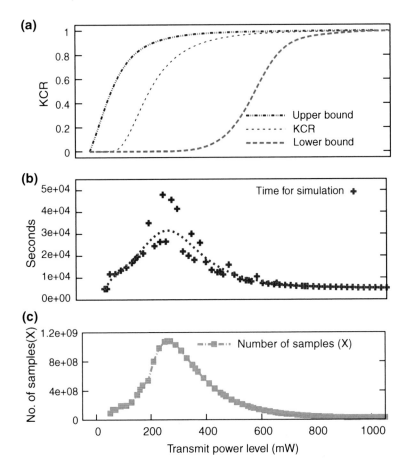

Fig. 6.18 Performance of CMCS method: **a** edge-packing bounds **b** simulation time **c** required number of samples

Its exact evaluation is known to be a hard problem (Harms 1995) and most of the current evaluation schemes depend on state generation methods.

For calculating performability of mesh networks, we set $F(\cdot)$ to be the network throughput. For a given traffic demand and network configuration (such as routing scheme, MAC protocol etc.), the mesh network yields a particular value of network throughput. Since traffic always flows between mesh nodes and gateways in a mesh network, we refer to its performability as k-center performability (KCP). The computation complexity of KCP problem is extremely high due to two reasons. First, as in all performability problems, generating 2^m states is a formidable combinatorial challenge. In case of KCP, the problem is especially difficult because the value of m can be very large for even a moderate sized network. The second reason is that even if there exists a method to efficiently generate network states, determining the throughput performance of a network state itself is a known hard problem (Jain et al. 2003).

Note that Monte Carlo state generation is especially inefficient (Harms 1995) in estimating $\bar{P}$ because it randomly generates states from the entire state space without any prior knowledge of the state's occurrence probability or performance. Due to this, other state generation methods such as most probable states generation (Gomes and Craveirinha 1998) or most performable states generation (James Jarvis 1996) are utilized for $\bar{P}$ estimation. The most probable states generation method generates states in order of their occurrence probability, and similarly the most performable states generation method generates states in order of their performance. We describe a method for KCP estimation that combines the advantages of both methods by generating states which have the most impact on KCP in terms of their state probability and performance. In the method, we first find a subset of edges that has the most definite impact on network performance. This subset is then input to most probable states generation method which yields KCP estimation. Before providing the details of the method, we first see how we can find the throughput performance of a given state of a mesh network efficiently.

6.4.1 Throughput Performance of Mesh Networks

There has been a large number of attempts (Brar et al. 2006; Gastpar and Vetterli 2002; Gupta and Kumar 2000; Jain et al. 2003; Jun and Sichitiu 2003) to precisely characterize the throughput capacity of a multihop wireless network given the traffic demand of nodes. The key limitations of these approaches are that either they yield an asymptotic performance or their calculation is computationally very expensive even for a moderate sized network. In our case, since the throughput has to be calculated for a large number of states, clearly such methods are not useful. But there is a key difference between the traffic characteristics of general ad-hoc networks and mesh access networks which allows a polynomial time calculation of capacity. Traffic in mesh access network always flows between mesh nodes and gateways as opposed to traffic between random pairs of nodes as in ad-hoc networks. This makes the calculation of the capacity more tractable because the capacity of the network clearly depends on how much aggregate traffic gateways can transfer.

In mesh networks, traffic aggregation near the gateways is always high which results in traffic bottlenecks near the gateways. The collision domains created around each gateway dictates how much traffic demand can actually be satisfied. The traffic accumulation around the gateways is so high that throughput reductions due to lower spatial reuse in other regions of the network has little or no effect on the overall capacity. This was shown by Pathak and Dutta (2011) for TDMA systems and by Robinson and Knightly (2007) and Robinson et al. (2008b) for CSMA-CA based medium access protocols.

Formally, let D bits per second be the total expected traffic demand of the mesh network. This is the cumulative demand of all mesh nodes in the network. Since every mesh node forwards its data to its gateway, let D_{g_i} denote the traffic demand associated with gateway g_i. Now, let CD_{g_i} be the collision domain of gateway g_i. The collision domain of a gateway is the largest set of links around the gateway such that every pair of links mutually interfere with each other. Let $|CD_{g_i}|$ denote the size of the collision domain, that is the total traffic of all the links in the collision domain. Now, as described by Pathak and Dutta (2011), Robinson et al. (2008b), $|CD_{g_i}|$ is the total traffic in a collision domain out of which D_{g_i} is the useful traffic. Meaning, during the time proportional to $|CD_{g_i}|$, the gateway is either transmitting, receiving or deferring for other transmission in the collision domain, while during the time proportional to D_{g_i}, the gateway is either transmitting or receiving (useful communications). This comparison yields a good estimation of how much wireless bandwidth a gateway utilizes for useful transmissions, which in turn determines the throughput capacity. The ratio $\delta_{g_i} = |CD_{g_i}|/D_{g_i}$ measures how poorly a gateway utilizes the wireless medium for useful transmission.

Due to high traffic accumulation near the gateways, the ratio δ_{g_i} is almost always larger than one. Now, if the capacity of the gateway g_i is B_{g_i} bits per second then the total aggregate throughput of the mesh network in bits per second is:

$$W = \sum_{i=1}^{k} \frac{B_{g_i}}{\delta_{g_i}} \qquad (6.23)$$

As described by Pathak and Dutta (2011), Robinson and Knightly (2007), the throughput calculation presented here assumes a fair MAC protocol, and an ideal routing and scheduling policy.

6.4.2 K-Center Performability

In this section, we describe how the most probable states generation method can be utilized for KCP evaluation. The method for calculating KCP consists of two phases. In the first phase, we identify a subset of edges which are most crucial in determining the KCP of a mesh network. This subset of edges becomes the input for most probable states method in the second phase. The method generates relatively fewer, but more important, states that contribute most towards the KCP value due to their occurrence probability and performance. We now discuss the details of these two phases.

6.4.2.1 Problem Size Reduction: Edge Selection

The purpose of edge selection phase is to identify l edges from the total m edges for which the most probable states (MPS) algorithm can be applied. In the states generated by the MPS algorithm, we fix the states of all edges other than these l

edges. Available computation power considerations can guide the choice of l. It is obvious that criteria for selecting l edges should depend on their expected impact on network performance and their existence probability. To capture these characteristics of edges, we divide the edge set E of $G_P(V, E, P)$ into three exhaustive and mutually exclusive subsets as below.

- X: an edge $e_i \in X$ if neither of the endpoints of e_i is a gateway and e_i does not interference with any gateway when active.
- Y: an edge $e_i \in Y$ if neither of the endpoints of e_i is a gateway and e_i interferes with a gateway when active.
- Z: an edge $e_i \in Z$ if one of the endpoints of e_i is a gateway and because of that e_i interferes with a gateway when active.

It is obvious that edges of Z will carry more traffic on average than edges of Y, and edges of Y will carry more traffic than edges of X in general. Since edges of Y and Z belong to a collision domain, their impact on network performance is also more substantial. Based on this understanding we determine the states of edges of each subset and also find l links for MPS algorithm as shown below.

- $\forall e_i \in X, s_{e_i} = 1$. This means that edges of set X are always in operating state in all network states that are generated by MPS algorithm. This is because these edges have the least impact on network performance as we saw in Sect. 6.4.1. Forcing these edges to be in operating mode allows maximum influx of traffic from all mesh nodes to gateways.
- $\forall e_i \in Y, s_{e_i} = 1$ if $p_{e_i} \geq 1 - p_{e_i}; s_{e_i} = 0$ otherwise. This means that edges in set Y are in their most probable state. As we know that these edges have a definite impact on network performance, we ensure that they remain in their most probable state in all network states generated by MPS algorithm, and affect the KCP evaluation only when they are likely to be in operational mode.
- In terms of KCP, the edges in Z are of foremost importance. Since the network throughput is most dependent on these edges, the choice of l links for MPS algorithm is made from this set. To do so, first we calculate probability weighted interference score I_{e_i} for every $e_i \in Z$. I_{e_i} is calculated as $I_{e_i} = p_{e_i} \times \Sigma_{e_j \in Z} \, p_{e_j}$, where $e_i, e_j \in Z$, and e_i and e_j mutually interfere. The I_{e_i} score of a link shows the likelihood that the link will be in operational mode and will interfere with other links from Z which are also probabilistically in their operational mode. We sort the links of Z in decreasing order of their I_{e_i} scores and create a set L using the first l links. These links, which have the most impact on network performance and state occurrence probability, are chosen for the MPS algorithm. Also, states of the remaining $Z - L$ links are fixed depending on the l links. Let $\rho = \max\{p_{e_i}, e_i \in L\}$. Now, for $e_i \in Z - L, s_{e_i} = 1$ if $p_{e_i} \geq \rho$, and $s_{e_i} = 0$ otherwise.

6.4.2.2 Most Probable States Algorithm

The MPS algorithm presented by Gomes and Craveirinha (1998) sequentially generates the most probable network states in order of decreasing probability. In this work, we utilize this algorithm with the l edges identified above. Since all other edges in $E - L$ have been assigned a fixed state, the MPS algorithm generates states of l edges in each iteration. In other words, the MPS algorithm behaves as if the network only contains edge set L and generates their states.

For each network state generated by the MPS algorithm, we find its throughput performance using the method proposed in Sect. 6.4.1. The MPS algorithm generates all $\chi = 2^l$ network states where state occurrence probability ($Pr(S)$) of each network state is determined using states of l links in the network state. Let $F(S)$ be the throughput performance of S. The performability estimate $\bar{P}$ is then calculated as $\bar{P} = \Sigma_{S \in \mathscr{P}} (F(S) \cdot Pr(S))$. For efficiency, we terminate the state generation when cumulative probability of generated network states reaches 0.99. As described by Gomes and Craveirinha (1998), the actual efficiency of MPS method depends on the average PQ factor. The average PQ factor is defined as the average of n_{e_i}/m_{e_i} ratio for all $e_i \in L$, where n_{e_i} is the probability that e_i is in its least probable state and $m_{e_i} = 1 - n_{e_i}$. Higher value of average PQ factor shows that the difference between link existence and failure probability is low for many links, and a higher number of states are required to be generated for cumulative state probability of 0.99. Lower values of average PQ factor, on the other hand, show that the first few most probable states are sufficient for a cumulative probability of 0.99.

6.4.2.3 Impact of Node Density and Transmission Power

As we did for KCA, we now apply the KCP evaluation method to study the impact of node density and transmission power on KCP. We apply the RF profile of GoogleWiFi ($\eta = 3.5$, $\sigma = 8\,dB$, $P_{min} = -80\,dBm$, link rate = 9 Mbps) as before. We inject 200 Mbps of traffic in the network which is equally divided among all mesh nodes. Figure 6.19 shows how KCP changes with varying node density and transmission power. For every density value, KCP initially increases, then decreases due to drops in network performance at higher power levels. It is further observed that for every density, KCP shows a sawtooth-wave like behavior with increasing transmission power. This can be explained as follows. After the first peak, with further increase in transmission power of nodes, performance generally decreases but the link existence probabilities increase (network availability increases). The decrease in the performance is due to interference. With increasing power, the set of interfering links remains stable for some interval, until a new set of links abruptly enter each others' interference range, at a certain power level. Thus performance displays a stepped behavior where it decreases sharply and then remains unchanged until another subsequent sharp decrease. This pattern of performance multiplied with increased availability results in intervals of slowly increasing performability, punctuated by sharp drops, in this region.

Note that the decrease in KCP is much faster for higher node densities, and with increasing density values, performance degradation is much sharper with increasing power level; this results into the gradual vanishing of the sawtooth-wave nature of KCP. As with KCA, the highest KCP (marked with circles in Fig. 6.19a–f) is achieved

Fig. 6.19 K-center Performability variation with transmission power and node density

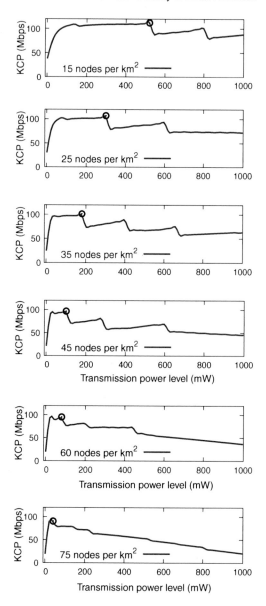

at lower power levels at higher node densities. However, absolute values of highest KCP are still comparable even at very high density values. This is because higher density usually yields a lower performance, but the availability with which it is delivered is higher.

For a given receiver sensitivity, links operate at a specific data rate in absence of any interference. Increasing receiver sensitivity increases link data rate and network

Fig. 6.20 Impact of receiver sensitivity on KCP

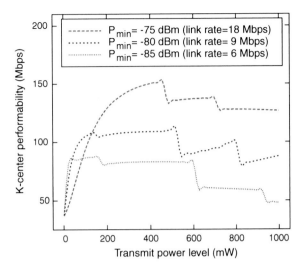

Fig. 6.21 Comparing KCP estimate to actual KCP value

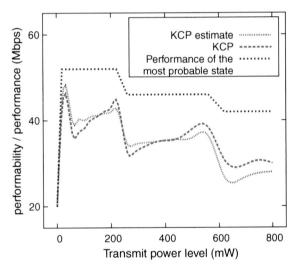

performance, but reduces the network availability. This motivates us to study the impact of receiver sensitivity on KCP. To do so, we fix the node density to be 15 nodes per km^2 and vary the transmission power and receiver sensitivity. We use the parameters of 802.11a for mapping receiver sensitivity to its corresponding data rate. While the variation of KCP with transmission power shows the same sawtooth pattern in each case, the absolute values do change. Figure 6.20 shows that in fact increasing receiver sensitivity results in increased KCP; this can be attributed to a sharper increase in performance, but a less sharp decrease in availability.

To examine the accuracy of the KCP evaluation method described, we apply it on a smaller network with 20 nodes, 40 links and 100 mbps of input traffic. We generate all ($2^{40} \approx 10^{12}$) network states and calculate the exact value of $\bar{P}$. We also apply the proposed KCP estimation method with $l = 20$. The results are shown in Fig. 6.21. It can be observed that the proposed method yields a good estimate of the actual KCP value. Also, the time requirement of proposed method is significantly lesser (nearly 26 h in worst case) than entire state space generation (approx. 198 h in worst case). Since the presented comparison is only for a small network, it is not certain that the same holds for larger networks, but it provides an indication that the method is useful. In the absence of any superior KCP estimation method for comparison, we also show the performance of the most probable state (in which all links are in their most probable state). This performance is what is typically considered when there is no consideration of link failures. It clearly demonstrates that while designing mesh networks in urban environments, performability (not absolute performance) should be considered in network planning.

We now consider the efficiency of the above KCP evaluation method. As mentioned before, the time requirement of the MPS algorithm depends on the average PQ factor. For a node density of 60 nodes per km^2, Fig. 6.22 shows the average PQ factor and the time required for KCP estimation. Higher values of average PQ factor indicates that more network states are required to be generated for a cumulative state probability of 0.99.

6.5 Designing for Availability and Performability

We have seen that KCA and KCP are both sensitive to design parameters that are well within the control of network engineers planning design and deployment of WMNs. For any fixed node density, KCA increases with increase in transmission power level. The increase is slower in cases of lower node density which shows that high transmission power level is necessary in order to maintain a high KCA in sparse deployments. The phase transition width decreases quickly with increase in node density. Also, node density and/or transmission power necessary to achieve a target KCA increases rapidly as target KCA requirement becomes more stringent (e.g. 0.999, 0.9999). This demonstrates that KCA calculation should be performed in network design phase to plan and provision a network for guaranteed KCA.

For a given density, KCP decreases with increase in transmission power due to decrease in throughput performance. The decrease is much sharper in case of higher node densities. It is generally understood from ad-hoc networks research that increase in density decreases the maximum achievable throughput performance. The same is not observed with performability since the maximum achievable KCP still remains comparable even at higher node densities. This is because higher node density reduces the performance but increases the network reliability on the other hand, which results in reasonable KCP values. Also, higher receiver sensitivity requirement guarantees higher KCP due to increase in network performance which is in contradiction to KCA results, where higher receiver sensitivity yields a lower KCA value.

Fig. 6.22 Efficiency of KCP
estimation method

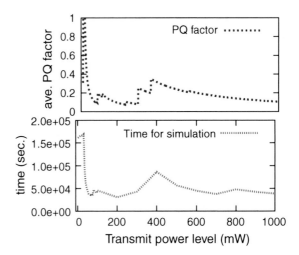

We believe that the concepts of KCA and KCP can be useful to network designers to plan and deploy networks, as can the evaluation methods we have described for them. Increasing the number of deployed nodes can result into unnecessary expenditure while increasing transmission power level might have performance related penalties. If a higher value of KCA is necessary (network must be highly available) while not guaranteeing any specific KCP, deploying the minimum number of nodes necessary for coverage, and utilizing high transmission power, can be a significantly more cost effective solution. Designing a network with a target KCA and KCP while meeting the coverage requirement is a challenging optimization problem. As an example, Fig. 6.19a, f show that for the RF profile under study, high values of both KCA and KCP can be achieved when node density is approximately 35 nodes per km^2 with a transmission power of approximately 200 mW. Further increase or decrease in node density or transmission power results into decrease of either KCA or KCP. This also shows that further investigation is required in identifying this unique node density and transmission power, that are specific to given network RF profiles.

6.5.1 Open Research Problems

So far we have discussed how to provide mesh network designers the tools which can help them in understanding the KCA and KCP of mesh networks. We now discuss how specific constrained design optimization tools can be built using the evaluation tools.

6.5.1.1 Network Deployment

Designing and deploying a mesh network with high KCA and KCP requires that link failures such as shadowing failures are taken into consideration from the planning phase. A typical network design and deployment problem requires a network to be deployed in the given geographical area such that necessary coverage is obtained. In any such problem, the following are usually available to network designers.

Potential locations. In the geographical area where network has to be deployed, potential locations are the locations where it is possible to place a mesh node. These locations are typically constrained by availability of buildings, street light poles, and other facilities in the neighborhood. This also includes a subset of locations for gateway nodes where wired connections (typically to the Internet) are feasible.

Environment R/F profile (RFP). It includes the wireless propagation characteristics of the outdoor environment where the network is being deployed. Using state-of-the-art measurement methods and site surveys, parameters such as signal path loss and shadowing coefficient can be determined. As we saw before, these are especially important in derivation of edge failure probabilities due to shadowing failures.

Mesh node profile (MNP). This is a set of parameters related to the radios used in mesh nodes. This includes information such as the level of noise, maximum transmit power level, receiver sensitivities and their corresponding data rates. In a two-tier mesh architecture, this information must be available for both backhaul and access tier radios. Also, MNP includes the information about the gain of the antenna (assumed to be omni-directional) used with backhaul tier radio. This together with maximum transmit power determines the effective transmit power of the radio.

Coverage constraint. The fundamental requirement of deploying a mesh network is to provide coverage of necessary locations. Hence, it is a common constraint to all our optimization problems. An accurate and practical way of finding the coverage requirements is to acquire all the locations where there is going to be some expected traffic demand. All such locations are then listed as a set of location which have to be covered by at least one mesh node. Since the access tier links (links between the clients and mesh nodes) are also prone to shadowing failures, the probability that received signal strength at a location from a mesh node is higher than the access tier receiver sensitivity threshold is also the probability that the location is covered by that mesh node. Let C be the set of all coverage locations. Now, the coverage requirement is described as follows. For all locations $c_i \in C$, the probability that c_i is covered by a mesh nodes should be at least θ_{c_i} and c_i should be covered by at least M_{c_i} mesh nodes. This representation allows consideration of locations which should be covered by multiple mesh nodes, because high traffic demand is expected at those locations.

Given this input information and constraint, a designer can intelligently choose the locations where mesh nodes should be deployed, and their transmit power levels, such that a target level of KCA and KCP is achieved, or KCA and KCP of the network is maximized. Here, the design and deployment involves determination of topological factors; especially node density, node positions and transmit power levels. Because there is a complex interplay between these topological factors and how they impact the KCA and KCP of the network, there are numerous flavors of the general design and deployment optimization problem. Below we articulate some of these problems and discuss how they can be useful.

- **Design problem 1**. Given potential locations, MNP, RFP, a set of mesh nodes and their transmit power levels, determine a node placement such that KCA and KCP are maximized while meeting the coverage constraint.
- **Design problem 2**. Given potential locations, MNP, RFP and a set of mesh nodes, determine a node placement and transmit power levels of mesh nodes such that KCA and KCP are maximized, and coverage constraint is met.
- **Design problem 3**. Given potential locations, MNP and RFP, determine a node placement and transmit power levels of mesh nodes such that the number of required mesh nodes are minimized, coverage constraint is met, and targeted KCA and KCP is achieved.

It is assumed in all the three problems that link failure probabilities are either available or derived using the shadowing failure model. Also, once a candidate node placement is found, its gateway placement is also necessary to calculate KCA and KCP. This problem of gateway assignment is well-studied; one of the current, more sophisticated methods is presented by Robinson et al. (2008b). This method assigns gateways to graph centers based on collision load balancing. In case of problem 1 and 2, if the given number of nodes can not satisfy the coverage, the conclusion is that there does not exist a valid and feasible node placement.

If we also assume that transmit power level of access tier radios are uniform for all mesh nodes (a reasonable practical case, even though this is not a requirement in any of the above mentioned design problems), it allows the following discussion with brevity.

In an optimal solution to Problem 1, nodes should be placed such that there is high link redundancy among the backhaul links and the lowest possible collisions around the gateways. The former guarantees higher KCA and latter provides a high value of KCP. With all possible locations and given mesh node power levels, there can be a large number of combinations where the coverage requirement is satisfied. Choosing a placement from all possible combinations which optimized the KCA and KCP can be a difficult problem. As a greedy heuristic solution to the problem, nodes can be placed in a square grid on the network area such that nodes with lower power levels are placed on grid locations nearer to high density coverage areas and nodes with higher power levels are placed on grid locations around low density coverage areas. The grid can then be perturbed such that nodes move towards their nearest coverage regions, allowing low power nodes to move more as compared to high power nodes. This results into areas where high coverage is available with lower collisions (higher KCP) while having sufficient redundancy in high power backhaul links (higher KCA). Of course, such greedy heuristics can result into unpredictable performance, but point the road to more sophisticated heuristics.

Similarly, if we assume that in Problem 2 the coverage locations are uniformly distributed in the network area, the solution to the problem of maximizing only the

KCA might be trivial. This is because placing the nodes as uniformly as possible (e.g. a square grid) and assigning them the maximum possible transmit power levels will always result in the highest KCA. However, the problem is less trivial in the case where coverage locations are not uniformly distributed in the network area. Further, when it is essential to maximize KCP along with KCA, increasing power levels to the maximum possible level might not yield a solution, since increased power levels can cause high interference and degraded performance. The difference between Problem 1 and 2 is that Problem 2 allows the choice of power levels along with the node placement. If we know the optimal solution to Problem 1, we can restate Problem 2 with an additional requirement that all the nodes should operate at the same power level. Once a candidate node placement is achieved, the power levels of nodes can be then modified to improve the KCP while maintaining the same KCA. Since this method of solving Problem 2 does not yield an optimal solution, this problem, too, remains open for more innovative solutions.

Problem 3 is different from the other two in that the objective is now to minimize the total number of required mesh nodes while meeting KCA, KCP and coverage constraints. This is especially an interesting problem from the designer's perspective, since it tries to minimize the cost associated with required number of mesh nodes in a greenfield deployment. As opposed to unrestricted coverage problems studied extensively in sensor networks, here the solution to the optimization problem guarantees target reliability and performability while minimizing the total cost. Methods such as restricted Steiner trees, where Steiner points are the mesh nodes and coverage locations are the leaf nodes, yield very poor solution since most such approaches try to minimize the number of backhaul link connections. This results in a sparse backhaul and a lower value of KCA and KCP. In Problem 3, it might be possible that even with a large number of nodes it is not possible to achieve a certain target KCP due to the inherent trade-off between the lower resilience of sparser topologies, and the lower throughput of denser topologies. In such a case, a solution to Problem 2 can yield the maximum possible value of KCA and KCP. When a designer can choose the node positions and transmit power levels for a new deployment, a solution to Problem 3 can indeed be very crucial in guaranteeing reliable performance.

6.5.1.2 Network Re-formation

The GoogleWiFi example motivates the consideration that if the network is already deployed, it may be possible to reform the network topology to achieve higher KCA and KCP. The design problems 1, 2 and 3 pertain mainly to greenfield deployments but related optimization problems can be defined for networks which are already deployed. For an existing mesh network, k-center availability and performability can be evaluated. If the network falls short in providing a reliable performance, it may be possible to refine its design (to some extent) to improve its KCA and KCP. We refer to this as the network re-formation problem.

Three ways of performing network reformation can be identified: adding additional mesh nodes, readjusting transmit power level of nodes, and repositioning the mesh nodes. Based on the network's current availability and performability, one or more out of these three options should be considered in re-formation.

Note that many of such network re-formation problems can be mapped back to problems of new network deployment with some modifications. As an example, the problem of repositioning of the mesh nodes is equivalent to design problem 1. The other two network reformation problems can be stated as below.

- **Design problem 4**. Given potential locations, MNP, RFP, node positions and power levels of currently deployed mesh nodes, determine the node placement and transmit power levels of additional mesh nodes such that number of required additional mesh nodes is minimized, and targeted KCA and KCP is achieved.
- **Design problem 5**. Given MNP, RFP and node positions of currently deployed mesh nodes, determine the power levels of mesh nodes such that KCA and KCP are maximized.

Note that the difference between Problem 3 and 4 is that Problem 4 pertains to the case where some of the mesh nodes are already deployed. We do not include the coverage constraint in both of these subproblems since an existing mesh network must already meet the coverage requirement and the re-formation is only necessary for improved KCA and KCP. If readjusting the power levels of mesh nodes (solution of Problem 5) attains a desired level KCA and KCP, additional nodes (Problem 4) might not be necessary in reformation. Since power control is clearly an inexpensive option for network reformation, solution to Problem 5 might be especially valuable. There can be similar other versions of design optimization problems in network reformation domain.

6.6 Conclusion

Wireless mesh networks may soon see deployment and use in a variety of service scenarios, including obvious ones such as community Internet access retrofit or municipal networks; the lessons learned in their design may also inform other usage models for multihop wireless networks that are not clearly visible now but will become so soon—perhaps vehicular infrastructure networks or mobile cloud computing. In maturing this paradigm to usefulness, its capability of providing continuous and predictable service, under fluctuating wireless conditions, will need to be understood well. The rich body of research in the design of such networks has provided tools and understanding that has well positioned the research community to address these challenges, and we expect to see significant activity in this area in the near future.

References

Al-Kuwaiti M, Kyriakopoulos N, Hussein S (2009) A comparative analysis of network dependability, fault-tolerance, reliability, security, and survivability. IEEE Commun Surv Tutor 11(2):106–124. doi:10.1109/SURV.2009.090208

Algirdas Avizienis BR, Laprie J-C (2001) Fundamental concepts of computer system dependability. In: IARP/IEEE-RAS workshop on robot dependability: technological challenge of dependable robots in human environments, Seoul, Korea

Annamalai A, Tellambura C, Bhargava V (2001) Simple and accurate methods for outage analysis in cellular mobile radio systems-a unified approach. IEEE Trans Commun 49:303–316. doi:10.1109/26.905889

Avizienis A, Laprie JC, Randell B, Landwehr C (2004) Basic concepts and taxonomy of dependable and secure computing. IEEE Trans Dependable Secur. Comput 1(1):11–33. doi:10.1109/TDSC.2004.2

Bela B (2001) Random graphs. Cambridge University Press, Cambridge Studies in Advanced Mathematics Series

Bettstetter C, Hartmann C (2005) Connectivity of wireless networks in a shadow fading environment. Wirel Netw 11:571–589

Brar G, Blough DM, Santi P (2006) Computationally efficient scheduling with the physical interference model for throughput improvement in wireless mesh networks. In: Proceedings of the 12th annual international conference on Mobile computing and networking (MobiCom '06), ACM, New York, NY, USA, pp 2–13. doi:10.1145/1161089.1161092

Bruno R, Conti M, Gregori E (2005) Mesh networks: commodity multihop ad hoc networks. IEEE Commun Mag 43(3):123–131. doi:10.1109/MCOM.2005.1404606

Camp J, Robinson J, Steger C, Knightly E (2006) Measurement driven deployment of a two-tier urban mesh access network. In: Proceedings of the 4th international conference on mobile systems, applications and services (MobiSys '06), ACM, New York, NY, USA, pp 96–109. doi:10.1145/1134680.1134691

Colbourn CJ (1987) The combinatorics of network reliability. Oxford University Press, New York

Egeland G, Engelstad P (2009) The availability and reliability of wireless multi-hop networks with stochastic link failures. IEEE J Sel Areas Commun 27(7):1025–1028

Fishman GS (1986) A monte carlo sampling plan for estimating network reliability. Oper Res 34(4):581–594

Gastpar M, Vetterli M (2002) On the capacity of wireless networks: the relay case. In: Proceedings of twenty-first annual joint conference of the IEEE computer and communications societies (INFOCOM 2002), vol 3. pp 1577–1586. doi:10.1109/INFCOM.2002.1019409

Gomes T, Craveirinha J (1998) Algorithm for sequential generation of states in failure-prone communication network. IEEE Proc-commun 145:73–79

GoogleWiFi (2007) Google WiFi access in Mountain View, California. http://wifigooglecom/

Grover WD (2003) Mesh-based survivable networks: options and strategies for optical, mpls, sonet and atm networking. Prentice Hall, Englewood Cliffs

Gupta P, Kumar P (2000) The capacity of wireless networks. IEEE Trans Inf Theory 46(2):388–404. doi:10.1109/18.825799

Harms DD (1995) Network reliability: experiments with a symbolic algebra environment. CRC Press, Boca Raton

Haverkort B, Marie R, Rubino GKT (2001) Performability modelling: techniques and tools. Wiley, New York

Hekmat R, Van Mieghem P (2003) Degree distribution and hopcount in wireless ad-hoc networks. In: The 11th IEEE international conference on networks (ICON2003), pp 603–609. doi:10.1109/ICON.2003.1266257

Jain K, Padhye J, Padmanabhan VN, Qiu L (2003) Impact of interference on multi-hop wireless network performance. In: Proceedings of the 9th annual international conference on mobile computing and networking (MobiCom '03), ACM, New York, NY, USA, pp 66–80. doi:10.1145/938985.938993

James Jarvis DRS (1996) An improved algorithm for approximating the performance of stochastic flow networks. Informs j comp 8:355–360

Jun J, Sichitiu M (2003) The nominal capacity of wireless mesh networks. IEEE Wirel Commun 10(5):8–14. doi:10.1109/MWC.2003.1241089

Kansas (2011) Kansas city—smart grid project. http://www.tropos.com/news/pressreleases/7_27_2011.php

Lee K, Lee HW, Modiano E (2011) Reliability in layered networks with random link failures. IEEE/ACM Trans Networking 19:1835–1848

Manzi E, Labbe M, Latouche FMG (2001) Fishman's sampling plan for computing network reliability. IEEE Trans Reliab 50:41–46

Pathak P, Dutta R (2011) Impact of power control on capacity of tdm-scheduled wireless mesh networks. In: IEEE international conference on communications (ICC), pp 1–6. doi:10.1109/icc.2011.5962410

Ponca (2007) Ponca city wireless network. http://wwwmyponcacitycom/wifi/poncaradioshtml

Rappaport TS (1996) Wireless communications: principles and practice. IEEE Press (The Institute of Electrical And Electronics Engineers)

Robinson J, Knightly E (2007) A performance study of deployment factors in wireless mesh networks. In: Proceedings of IEEE INFOCOM 2007, Anchorage

Robinson J, Swaminathan R, Knightly EW (2008a) Assessment of urban-scale wireless networks with a small number of measurements. In: Proceedings of the 14th ACM international Conference on Mobile Computing and Networking (Mobicom'08)

Robinson J, Uysal M, Swami R, Knightly E (2008b) Adding capacity points to a wireless mesh network using local search. In: Proceedings of the IEEE INFOCOM 2008

Robinson J, Singh M, Swaminathan R, Knightly E (2010) Deploying mesh nodes under non-uniform propagation. In: Proceedings of IEEE INFOCOM 2010, pp 1–9. doi:10.1109/INFCOM.2010.5462038

Roskind J, Tarjan R (1985) A note on finding minimum-cost edge-disjoint spanning trees. Math Oper Res 10:701–708

Tutte WT (1961) On the problem of decomposing a graph into n connected factors. J Lond Math Soc 36(1):221–230

Vos E (2011) Okmulgee, Oklahoma deploys municipal wi-fi network for public safety, government use. http://www.muniwireless.com/2011/06/16/okmulgee-oklahoma-deploys-municipal-wifi-network

Xing F, Wang W (2008) On the critical phase transition time of wireless multi-hop networks with random failures. In: Proceedings of the 14th ACM international conference on mobile computing and networking

Index

2-phase MAC protocol, 75
802.11e, 151

A

Access, 11
Additive increase multiplicative decrease
      (AIMD), 112
Additive interference, 42
Additive physical interference model, 66, 139
Ad-hoc network, 4
Ad hoc on-demand distance vector (AODV),
      100
Aggregate MAC service data unit, 19
A-MSDU, 19
Applications, 7
Asynchronous interference, 48
Availability, 173
    2-terminal, 174
    k-center, 184
    k-terminal, 174
    mesh network, 185
    network, 171
    sustainable energy systems, 142
Availability polynomial, 176

B

Back-pressure framework, 147, 162
    cross-layer protocol, 157
    CSMA/CA, 150
    rate control, 149
    routing, 157
    scheduling, 148
    XPRESS implementation, 160

Backhaul, 11, 13
Balanced incomplete block design, 80
Beam-forming, 17, 43
Betterapproach to mobile ad-hoc
      networking (BATMAN), 101
Bottleneck collision domain, 58, 201
Broadcast, 107

C

Cellular relay, 20
Channel assignment, 78
    and routing, 137
    and routing and scheduling, 138
    and scheduling, 138
    dynamic, 78
    static, 78
Channel switching cost (CSC), 98
Clear-to-send-simultaneously, 74
Clustered topology, 56
CLUSTERPOW, 51
Co-channel interference, 82
CogMesh, 28
Cognitive medium access control, 122
Cognitive mesh networks, 28, 116
      spectrum sensing and congestion
          control, 161
      spectrum sensing and routing, 160
Cognitive radio networks, 28
Common power level (COMPOW), 51
Communication range, 40
Community mesh networks, 179
Community wireless networks, 4
Conditional state generation, 192
Cone-based topology control, 62

P. H. Pathak and R. Dutta, *Designing for Network and Service Continuity in Wireless Mesh Networks*, Signals and Communication Technology, DOI: 10.1007/978-1-4614-4627-9, © Springer Science+Business Media New York 2013